Science and Technology Concepts–Secondary™

Exploring
Planetary
Systems

Student Guide

National Science Resources Center

The National Science Resources Center (NSRC) is operated by the Smithsonian Institution to improve the teaching of science in the nation's schools. The NSRC disseminates information about exemplary teaching resources, develops curriculum materials, and conducts outreach programs of leadership development and technical assistance to help school districts implement inquiry-centered science programs.

Smithsonian Institution

The Smithsonian Institution was created by an act of Congress in 1846 "for the increase and diffusion of knowledge..." This independent federal establishment is the world's largest museum complex and is responsible for public and scholarly activities, exhibitions, and research projects nationwide and overseas. Among the objectives of the Smithsonian is the application of its unique resources to enhance elementary and secondary education.

STC Program™ Project Sponsors

National Science Foundation

Bristol-Meyers Squibb Foundation

Dow Chemical Company

DuPont Company

Hewlett-Packard Company

The Robert Wood Johnson Foundation

Carolina Biological Supply Company

Science and Technology Concepts–Secondary™

Exploring
Planetary
Systems

Student Guide

The **STC** *Program*™

Smithsonian Institution
National Science Resources Center

www.carolinacurriculum.com

Published by Carolina Biological Supply Company
Burlington, North Carolina

NOTICE This material is based upon work supported by the National Science Foundation under Grant No. ESI-9618091. Any opinions, findings, and conclusions or recommendations expressed in this material are those of the authors and do not necessarily reflect views of the National Science Foundation or the Smithsonian Institution.

This project was supported, in part, by the **National Science Foundation**.
Opinions expressed are those of the authors and not necessarily those of the foundation.

ISBN 978-1-4350-0669-0

Published by Carolina Biological Supply Company, 2700 York Road, Burlington, NC 27215.
Call toll free 1-800-334-5551.

1406

Science and Technology Concepts—Secondary™
Exploring Planetary Systems

The following revision was based on the STC/MS™ module *Earth in Space.*

Developer
Dane J. Toler

Scientific Reviewer
Ian MacGregor
Senior Scientist
National Science Resources Center

Illustrator
John Norton

**Developer/Writer
Interactive Whiteboard Activities**
Sandy Ledwell, Ed.D

Writers/Editors
Amy Charles
Ian Mark Brooks
Devin Reese

Photo Research
Jane Martin
Devin Reese

National Science Resources Center Staff

Executive Director
Thomas Emrick

Program Specialist/Revision Manager
Elizabeth Klemick

Contractor, Curriculum Research and Development
Devin Reese

Publications Graphics Specialist
Heidi M. Kupke

Carolina Biological Supply Company Staff

Director of Product and Development
Cindy Morgan

Marketing Manager, STC-Secondary™
Jeff Frates

Curriculum Editors
Lauren Goldsmith
Gary Metheny

Managing Editor, Curriculum Materials
Cindy Vines Bright

Publications Designers
Trey Foster
Charles Thacker
Greg Willette

Science and Technology Concepts for Middle Schools™
Earth in Space
Original Publication

Module Development Staff

Developer/Writer
Carol O'Donnell

Science Advisors

Stanley Doore, Meteorologist (retired)
National Weather Service
National Oceanic and Atmospheric
Administration

Brian Huber, Micropaleobiologist
Department of Paleobiology
National Museum of Natural History
Smithsonian Institution

Ian MacGregor, Director (retired)
Division of Earth Sciences
National Science Foundation

David Williams, Planetary Scientist
National Space Science Data Center
NASA Goddard Space Flight Center

James Zimbelman, Planetary Geologist
Center for Earth and Planetary Studies
National Air and Space Museum
Smithsonian Institution

Contributing Writers
Phil Berardelli
Lynda DeWitt
Carolyn Hanson
Linda Harteker
Scott Paton
Catherine Stephens

Illustrator
Max-Karl Winkler

STC/MS™ Project Staff

Principal Investigator
Sally Goetz Shuler, Executive Director, NSRC

Project Director
Kitty Lou Smith

Curriculum Developers
David Marsland
Henry Milne
Carol O'Donnell
Dane J. Toler

Illustration Coordinator
Max-Karl Winkler

Photo Editor
Christine Hauser

Graphic Designer
Heidi M. Kupke

STC/MS™ Project Advisors

Tom Albert, Teacher-in-Residence, NASA Goddard Space Flight Center

Cassandra Coombs, Director, NASA Southeast Regional Clearinghouse, College of Charleston

Stanley Doore, Meteorologist (retired), National Weather Service, National Oceanic and Atmospheric Administration

Ann Dorr, Teacher (retired), Fairfax County, Virginia, Public Schools; Board Member, Minerals Information Institute

Andrew Fraknoi, Astronomical Society of the Pacific; Professor, Department of Astronomy, Foothills College

Jackie Faillace Getgood, Supervisor of Mathematics, Spotsylvania County, Virginia, Public Schools

Marvin Grossman, Associate Director, Project ARIES, Harvard University, Harvard-Smithsonian Center for Astrophysics

Patricia Hagan, Middle School Science Specialist, Montgomery County, Maryland, Public Schools

Matthew Holman, Astrophysicist, Harvard-Smithsonian Center for Astrophysics

Brian Huber, Micropaleobiologist, Department of Paleobiology, National Museum of Natural History, Smithsonian Institution

Ian MacGregor, Director (retired), Division of Earth Sciences, National Science Foundation

Brian Marsden, Senior Astrophysicist, Associate Director, Smithsonian Astrophysical Observatory, Harvard-Smithsonian Center for Astrophysics

Timothy McCoy, Meteorite Specialist, Department of Mineral Sciences, National Museum of History, Smithsonian Institution

Stephanie Stockman, Planetary Geologist, Science Systems and Applications, Inc.; NASA Goddard Space Flight Center

David Williams, Planetary Scientist, National Space Science Data Center, NASA Goddard Space Flight Center

James Zimbelman, Planetary Geologist, Center for Earth and Planetary Studies, National Air and Space Museum, Smithsonian Institution

Acknowledgments

The National Science Resources Center gratefully acknowledges
the following individuals and school systems for their assistance
with the national field-testing of *Earth in Space:*

Bozeman Public School District
Bozeman, Montana

Site Coordinator
Myra Miller, Keystone Project Director

Sheri Konietzko, Teacher
Sacajawea Middle School

Ana Morris, Teacher
Sacajawea Middle School

Joann Watson, Teacher
Chief Joseph Middle School

Anderson Oconee Pickens Hub
Clemson, South Carolina

Site Coordinator
Elizabeth Edmondson

David Pepper, Teacher
Seneca Middle School, Seneca

Alan Weekes, Teacher
Pickens Middle School, Pickens

Ali Wienke, Teacher
Wren Middle School, Piedmont

Fort Bend Independent School District
Missouri City, Texas

Site Coordinator
Mary Ingle, Director of Secondary Science

Tom Grubbs, Teacher
Lake Olympia Middle School

Kirlew Matthie, Teacher
Lake Olympia Middle School

Scott McKie, Teacher
Lake Olympia Middle School

Schenectady City School District
Schenectady, New York

Site Coordinator
Arden Rauch

Claire Godlewski, Teacher
Oneida Middle School

Danielle Hartkern, Teacher
Central Park Middle School

Ed Pfeifer, Teacher
Schenectady High School

Spotsylvania County School District
Spotsylvania, Virginia

Site Coordinator
Katie Wallet, Supervisor of Science

Mary Hardy, Teacher
Ni River Middle School

The NSRC thanks the following individuals for their assistance during the development of *Earth in Space*:

Dennis Schatz, Associate Director
Pacific Science Center, Seattle, Washington

Rose Steinet, Photo Librarian
Center for Earth and Planetary Studies, National Air and Space Museum
Smithsonian Institution, Washington, DC

The NSRC appreciates the contribution of its
STC/MS project evaluation consultants—

Center for the Study of Testing, Evaluation,
and Education Policy (CSTEEP), Boston College

Joseph Pedulla
Director, CSTEEP

Preface

Community leaders and state and local school officials across the country are recognizing the need to implement science education programs consistent with the National Science Education Standards to attain the important national goal of scientific literacy for all students in the 21st century. The Standards present a bold vision of science education. They identify what students at various levels should know and be able to do. They also emphasize the importance of transforming the science curriculum to enable students to engage actively in scientific inquiry as a way to develop conceptual understanding as well as problem-solving skills.

The development of effective standards-based, inquiry-centered curriculum materials is a key step in achieving scientific literacy. The National Science Resources Center (NSRC) has responded to this challenge through Science and Technology Concepts–Secondary™. Prior to the development of these materials, there were very few science curriculum resources for secondary students that embodied scientific inquiry and hands-on learning. With the publication of STC-Secondary™, schools will have a rich set of curriculum resources to fill this need.

Since its founding in 1985, the NSRC has made many significant contributions to the goal of achieving scientific literacy for all students. In addition to developing Science and Technology Concepts–Elementary™—an inquiry-centered science curriculum for grades K through 6—the NSRC has been active in disseminating information on science teaching resources, preparing school district leaders to spearhead science education reform, and providing technical assistance to school districts. These programs have had a significant impact on science education throughout the country. The transformation of science education is a challenging task that will continue to require the kind of strategic thinking and insistence on excellence that the NSRC has demonstrated in all of its curriculum development and outreach programs. The Smithsonian Institution, our sponsoring organization, takes great pride in the publication of this exciting new science program for secondary students.

Letter to the Students

Smithsonian Institution
National Science Resources Center

Dear Student,

The National Science Resources Center's (NSRC) mission is to improve the learning and teaching of science for K-12 students. As an organization of the Smithsonian Institution, the NSRC is dedicated to the establishment of effective science programs for all students. To contribute to that goal, the NSRC has developed and published two comprehensive, research-based science curriculum programs: Science and Technology Concepts-Elementary™ and Science and Technology Concepts-Secondary™.

By using the STC-Secondary™ curriculum materials, we know that you will build an understanding of important concepts in life, earth, and physical sciences; learn critical-thinking skills; and develop positive attitudes toward science and technology. The National Science Education Standards state that all secondary students "...should be provided opportunities to engage in full and partial inquiries.... With an appropriate curriculum and adequate instruction, ... students can develop the skills of investigation and the understanding that scientific inquiry is guided by knowledge, observations, ideas, and questions."

STC-Secondary also addresses the national technology standards published by the International Technology Education Association. Informed by research and guided by standards, the design of the STC-Secondary units address four critical goals:

• Use of effective student and teacher assessment strategies to improve learning and teaching

• Integration of literacy into the learning of science by giving students the lens of language to focus and clarify their thinking and activities

• Enhanced learning using new technologies to help students visualize processes and relationships that are normally invisible or difficult to understand

• Incorporation of strategies to actively engage parents to support the learning process

We hope that by using the STC-Secondary curriculum you will expand your interest, curiosity, and understanding about the world around you. We welcome comments from students and teachers about their experiences with the STC-Secondary program materials.

Thomas Emrick
Executive Director
National Science Resources Center

Navigating an STC–Secondary™ Student Guide

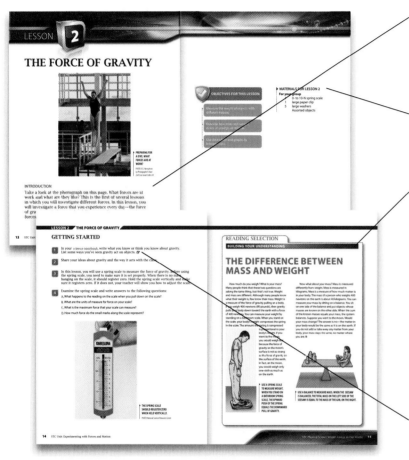

INTRODUCTION
This short paragraph helps get you interested about the upcoming inquiries.

MATERIALS
This helps you get organized and prepare for your inquiries.

READING SELECTION:
BUILDING YOUR UNDERSTANDING
These reading selections are part of the lesson, and give you information about the topic or concept you are exploring.

NOTEBOOK ICON ✎
During the course of an inquiry, you'll record data in different ways. This icon lets you know to record in your science notebook. Student sheets are called out when you're to write there. You may go back and forth between your notebook and a student sheet. Watch carefully for the icon throughout the procedure.

SAFETY TIPS
Safety in the science classroom is very important. Tips throughout the student guide will help you to practice safe techniques while conducting investigations. It is very important to read and follow all safety tips.

PROCEDURE
This tells you what to do. Sometimes the steps are very specific, and sometimes they guide you to come up with your own investigation and ways to record data.

REFLECTING ON WHAT YOU'VE DONE
These questions help you think about what you've learned during the lesson's inquiries, apply them to different situations, and generate new questions. Often you'll discuss your ideas with the class.

READING SELECTION: EXTENDING YOUR KNOWLEDGE
These reading selections come after the lesson, and show new ways that the topic or concept you learned about during the lesson can be applied, often in real-world situations.

GLOSSARY
Here you can find scientific terms defined.

INDEX
Locate specific information within the student guide using the index.

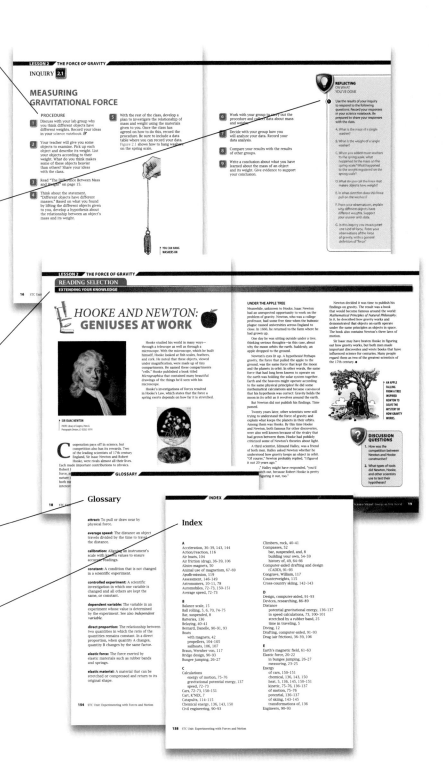

Contents

CONTENTS

THINKING ABOUT OUR SOLAR SYSTEM

INTRODUCTION

By looking at the sky, you have made direct observations of the Sun and Moon. But careful observation of the nighttime sky reveals that there are other objects to observe and study. One set of objects is the "wanderers," or stars that move against the background of more distant stars. The ancients called these wanderers "planets." Over the centuries, curiosity about the planets has led scientists and astronomers to develop and use tools to learn more about the planets (and other objects) that comprise the solar system.

Early astronomers used their eyes alone to make direct observations of the Sun, Moon, and planets. But most of what astronomers know about our solar system comes from the use of tools that reveal details that cannot be seen with the eyes alone.

▶ THIS COMPOSITE SHOWS SEVERAL IMAGES OF THE PLANETS THAT WERE TAKEN BY VARIOUS SPACECRAFT AND MADE INTO ONE PHOTO.

PHOTO: NASA Jet Propulsion Laboratory

This unit offers an opportunity to explore the solar system and learn about the objects in it. You will compare planets' sizes and distances from each other, investigate their surface features, learn about how they move, and what holds the solar system together. You will also learn about other objects in the solar system.

Let's think about what you already know about the solar system and identify some things you would like to learn.

OBJECTIVES FOR THIS LESSON

Record your ideas and questions about the solar system.

Record your responses to 10 common questions about the solar system.

Analyze the class's responses to these 10 questions.

▶ MATERIALS FOR LESSON 1

For you

10   self-stick notes

GETTING STARTED

1 Read the Introduction and Objectives for this lesson. (Try to do this before every lesson.)

2 Record in your science notebook 5-10 things you know about the solar system. ✑

3 Share what you know about the solar system with the class.

INQUIRY 1.1

EXAMINING OUR IDEAS ABOUT THE SOLAR SYSTEM

PROCEDURE

1 Answer each of the following 10 questions individually in your science notebook. Label each answer with the corresponding letter (A-J). ✑

A. What tools do we have for exploring our solar system?

B. What makes up our solar system?

C. Look at Figure 1.1. What surface feature is shown? Where would you find this surface feature in the solar system?

D. If you could travel to another planet, what would happen to your weight? Explain your reasoning.

E. Look at Figure 1.2. What processes created this landform? Does this landform exist on other planets or moons? Explain why or why not.

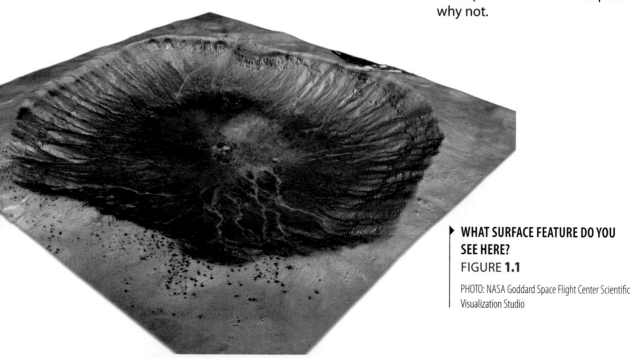

▶ **WHAT SURFACE FEATURE DO YOU SEE HERE?**
FIGURE 1.1

PHOTO: NASA Goddard Space Flight Center Scientific Visualization Studio

▶ **WHAT PROCESSES CREATED
THIS LANDFORM?**
FIGURE **1.2**

PHOTO: J.D. Griggs, U.S. Geological Survey

Inquiry 1.1 continued

F. Where does gravity exist? Where is gravity strongest? Where is it weakest? Why?

G. What keeps planets, comets, asteroids, and dwarf planets in orbit around the Sun?

H. Look at Figure 1.3. What do you think this object is? Where would you find it in our solar system?

I. Look at Figure 1.4. What do you see in the night sky besides stars? Write what you know about this object.

J. What are fossils? What do fossils reveal about Earth?

▶ **WHAT DO YOU THINK THIS OBJECT IS?**
FIGURE **1.3**

PHOTO: NASA Jet Propulsion Laboratory

▶ **WHAT DO YOU SEE IN THE NIGHT SKY?**
FIGURE **1.4**

PHOTO: Brian McLeod

2 Record each of your answers on a separate self-stick note. Write the matching letter (A–J) in the corner of the note.

3 Your teacher will circulate folders among members of your group. Each folder contains a question (A–J) and a photograph. Examine the photograph and read its matching question. Post your self-stick answers inside each matching folder.

4 Once you have posted all your answers to Questions A–J, your group will get one or two completed folders on which to report. Read all the posted answers on each completed folder. Put all the answers that are the same, or nearly the same, together in piles.

5 From each pile, select one answer and post it on the inside of the folder. On that answer, indicate how many times other students gave it as an answer. Put aside the duplicates.

6 Post any unique or original answers inside the folder.

7 When all groups are ready, report your findings to the class. Ask your classmates if they have any questions or want to debate any of the statements. Be prepared to revisit these statements throughout the unit.

REFLECTING
ON WHAT
YOU'VE DONE

1 Discuss the following questions with your class.

A. Are any of the questions (A–J) you answered during Inquiry 1.1 related to the same topic? Explain your answer by giving an example.

B. What can we learn about Earth by studying the solar system?

C. How is Earth different from other planets? How is Earth similar to other planets?

2 Record in your science notebook any questions you have about the solar system in general. Label these questions "What I Want to Learn About the Solar System."

3 Share your questions in a class discussion. Your teacher will record your ideas. You will try to answer these questions as you work through this unit.

4 In your science notebook, summarize what you learned (or did) in this lesson. Date your entry.

Astronomy: Looking Back

What do you see when you look at the night sky? Your first answer may be "stars." Almost everything we see at night looks like a point of light. But did you know that those points of light also include planets, moons around other planets, meteors, asteroids, and even comets? All these heavenly bodies and our Sun—a star itself—are part of the cosmic neighborhood that we call the solar system.

How did we come to learn so much about the solar system? Today, astronomers use technological advances such as orbiting space telescopes, satellites, and space probes to take pictures of the solar system and deep space. However, while modern astronomy—the science of observing the sky—is only a few centuries old, skywatching has been a pastime throughout history.

ANCIENT SKYWATCHING

During ancient times, people could easily see the Sun and Moon, five bright wandering stars (planets), an occasional comet, and frequent meteors. But they did not know what these objects were, so they made up stories to describe what they saw. To the ancients, everything in the night sky was magical, and the stories that they told, called myths, explained that magic. Astrology grew out of these early myths. (Astrology is pseudo-science and is the belief that events in the sky—the "heavens"—control our lives and predict the future.)

The ancients began to identify patterns in the sky. These early skywatchers recognized patterns in the stars' positions, gave those patterns names, and told stories to explain how these constellation "pictures" came to be. They

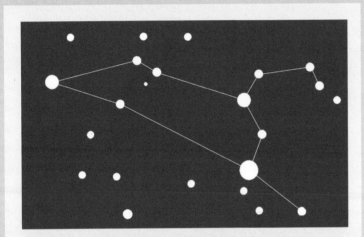

CONSTELLATION LEO, SHAPED LIKE A LION, IS BEST SEEN IN EARLY SPRING WHEN IT IS HIGH IN THE SKY IN THE NORTHERN HEMISPHERE. REGULUS IS THE BRIGHT STAR THAT MARKS THE LION'S HEART.

also observed the Sun as it appeared to move across the daytime sky and watched as the Moon seemed to follow a similar path.

The earliest skywatchers who kept records were the Sumerians, Babylonians, Egyptians, and Chinese. Their records date as far back as 5000 BC. The Greeks, in 1000 BC, continued to try to find order in the sky and in the motions of objects in the sky. They went beyond earlier, simple observations and tried to develop theories or models for the nature of celestial objects and their motions.

CLASSIFYING "STARS"

The ancients thought that everything in the sky was a star. They classified stars this way:

- "Fixed stars" that didn't change and didn't move relative to one another (which we know as stars today)
- "Shooting stars" that flashed across the sky (these are now known as meteors)
- "Hairy stars" that moved across the sky with a tail following behind them (we know these as comets)
- "Wandering stars" that moved across the sky following the Sun's path (these are now known as planets)

The Greeks gave the name *planetes*, meaning "wanderers," to the Sun, Moon, and planets—which to them were points of light that moved against the background of fixed

READING SELECTION
EXTENDING YOUR KNOWLEDGE

stars. It became apparent over time that these *planetes* were part of a closely associated group of objects that we now know as our solar system.

EARTH-CENTERED SYSTEM

The Greeks believed that Earth was the center of the universe and that all the planets and stars moved around Earth. According to Greek philosophy, everything about Earth was perfect. The Greeks believed that the sphere was the perfect shape, and they theorized that the planets, the Sun, and the Moon were attached to huge, turning, transparent crystalline spheres that were centered on Earth. Their model was geocentric, which means Earth-centered.

In the second century AD in Alexandria, Egypt, the Greek astronomer Claudius Ptolemy wrote a book called *The Almagest*. In this book, Ptolemy described a model of the heavens in which planets moved in little circles that moved on bigger circles around Earth. His geocentric model with circles on circles explained why the planets appeared sometimes to move backward (retrograde) in the sky. For almost 1400 years, people believed Ptolemy's Earth-centered model to be correct.

SUN-CENTERED SYSTEM

In the early 1500s, the astronomer Nicolaus Copernicus modified Ptolemy's geocentric model with the revolutionary theory that the Sun, not Earth, was the center of the planetary system and universe. Copernicus's model was a Sun-centered, or heliocentric, model. Some people who heard of Copernicus's heliocentric model thought it was right, but they needed data to support it. After Copernicus died in 1543, the Danish astronomer Tycho Brahe set out to make accurate observations of the planets and stars to determine which model,

Earth-centered or Sun-centered, was correct. Brahe came to believe that the planets seen in the sky orbited the Sun. But he still believed that the Moon and the Sun revolved around Earth. After his death, Brahe's assistant, Johannes Kepler, used Brahe's data for Mars to develop a Sun-centered model in which all the planets—including Earth—orbited the Sun. Kepler's model also accurately described and predicted the motions of the planets in the sky.

The invention of the telescope by Hans Lippershey in 1608 changed astronomy forever. This new invention made distant objects appear to be nearer and clearer. The Italian scientist Galileo Galilei was the first to adapt one of these telescopes for astronomical use in 1610. Galileo's telescope made fainter objects visible, and enabled him to see details that could not be seen with his eyes alone. He discovered new things about the Moon and planets. Among Galileo's first discoveries were four moons revolving around Jupiter. Since Galileo could see for himself that these moons orbited Jupiter, he could demonstrate that Earth was not the center of all motion in the universe. Galileo was arrested for his discoveries and claims, which contradicted many beliefs held by the Church. However, Galileo and many others made new discoveries over the years, including the rings of Saturn, sunspots, and other phenomena.

LOOKING BEYOND

Today, we see and study the same sky that the skywatchers of old saw and studied. The biggest difference is that we have the benefit of technology and much more knowledge to explain our observations. Exciting new discoveries about the universe are made every day. ■

▶ GALILEO USED ONE OF THE FIRST TELESCOPES TO LOOK AT THE SKY.

DISCUSSION QUESTIONS

1. What is the difference between astronomy and astrology?

2. What theories about the solar system have been disproved through scientific observation?

Science Fiction —Science Fact

What kinds of books do you like to read? Many people turn to science fiction. Adults may enjoy writers such as Isaac Asimov and Arthur Clarke, whose works are modern classics. Younger people may prefer a writer like Madeline L'Engle, who wrote *A Wrinkle in Time* and *A Swiftly Tilting Planet*.

Why is science fiction so popular? One reason is that science fiction tells a good story that removes us from our ordinary surroundings. Authors who make good use of their imaginations can create convincing and inventive worlds.

But science fiction is different from fantasy—it is not based on imagination alone. Real science fiction is based on scientific principles and a solid understanding of the real world and how it works. Some of the best-known authors of science fiction are scientists themselves. Asimov had a Ph.D. in chemistry and he taught in a medical school—whenever he wasn't writing one of his hundreds of books!

Asimov defined science fiction as writing that is concerned with the "impact of scientific advance on human beings." He thought science fiction was a good way to make science more accessible to everyone.

▶ **JULES VERNE'S BOOK** *FROM THE EARTH TO THE MOON*
PHOTO: Courtesy of Smithsonian Institution Libraries, National Air and Space Museum, Washington, DC

Once you know what science fiction is, you can see how something that was once featured in science fiction can become science fact. In 1865, the French author Jules Verne wrote a book, *From the Earth to the Moon*, that described a manned voyage to the Moon. Most French people at that time were probably skeptical—*"C'est impossible!"* they may have said.

But just about 100 years after Jules Verne imagined a trip to the Moon, U.S. astronaut Neil Armstrong and his team landed the spacecraft *Eagle* on the Moon's surface. Other works by Jules Verne describe underwater vessels—what we now call submarines—and even something similar to today's television sets decades before these things were invented.

▶ AUTHOR JULES VERNE SURROUNDED BY ILLUSTRATIONS OF SOME OF HIS IDEAS

PHOTO: Courtesy of Smithsonian Institution Libraries, National Air and Space Museum, Washington, DC

READING SELECTION
EXTENDING YOUR KNOWLEDGE

Science fantasy can be great reading. One of the most famous books of the late 1800s was *War of the Worlds* by H.G. Wells. In that story, Martians invade Earth. A radio broadcast based on the book aired on Halloween night in 1938 and caused millions of people to believe that an invasion from Mars was actually taking place. Some people even decided to evacuate their cities!

▶ SOME CITIZENS FLED THEIR HOMES IN PANIC AFTER HEARING THE RADIO BROADCAST "WAR OF THE WORLDS."

Scientists doubted Wells's fantastic theory about Martians. And when the spacecraft *Viking 1* landed on Mars in 1976, people learned that this famous author had indeed been writing science *fantasy* and not science *fiction*. There were no signs of life on the planet. Wells's book is good reading, but his tales are not necessarily based on true science.

Science fiction helps shape our vision of how we will live tomorrow. If something can be imagined, has practical applications, and is scientifically possible, chances are it will become a reality. Jules Verne, one of the greatest science fiction writers of all time, was able to imagine the future of space travel.

Why don't you give it a try? Let your imagination soar. Can you think of something that is science fiction today—but may become science fact tomorrow? ∎

DISCUSSION QUESTIONS

1. Some science fiction fans make a game out of finding scientific inaccuracies in stories and TV shows. Is it really important that science fiction get the science right? Why or why not?

2. What processes do scientists use to distinguish fiction from fact?

Alike, Yet Different
The Planets in Perspective

▶ THIS IMAGE, CONSTRUCTED USING DATA ON SATURN'S RINGS, USES COLORS TO SHOW THE VARIOUS PARTICLE SIZES OF THE SWIRLING MATERIALS (PURPLE COLOR INDICATES AREAS WITH THE LARGEST PARTICLES).

PHOTO: NASA Jet Propulsion Laboratory

What makes a planet a planet? It is of a certain size, it orbits a star (for our solar system, the Sun of course), and it isn't luminous—in other words, it doesn't make its own light. These defining characteristics are shared by all eight of the planets in our solar system. Yet each planet is distinct enough that in a game of "Guess the Planet," just a few clues would allow you to identify one from another.

The game's first question might be, Is it a terrestrial planet? The root of terrestrial is "terra-," or "earth": a terrestrial planet is earthlike. Four planets in our solar system—Mercury, Venus, Earth, and Mars—are terrestrial, meaning that they have a lot of exposed solid surface and relatively thin atmospheres. These terrestrial planets at one time had thicker atmospheres, but lost them during a period when the Sun generated huge explosions. The explosions produced such intense solar radiation that it blew off the atmospheres of the four innermost planets. So, these planets are now small and rocky.

How can Venus be considered small when it is 12,000 kilometers (about 7500 miles) in diameter, roughly equivalent to 1,000,000 school buses, across? Well, it's small compared to the outer four planets, Jupiter, Saturn, Uranus, and Neptune, which are called the gaseous, or Jovian, giants. As their name implies, they're enormous and they're made mostly of gases. Saturn's diameter, for instance, is ten times that of Venus. All four rotate at very high speeds. Although Saturn's rings are the easiest for us to see, the Jovian planets all have rings around them, composed of swirling rock, planetary dust, and ice shards.

WHERE IS ALL THE WATER?

Earth is the only planet that currently has large bodies of liquid water on it: about 1.25 billion cubic kilometers (nearly 300 million cubic miles) of it. Earth's vast oceans set it apart from all the other planets. A narrow planetary temperature range keeps most of Earth's water in liquid form. Water continually evaporates to create rainfall that waters the lush vegetation on the continental land masses. While other terrestrial planets probably had surface water in the past, for various reasons they are dry now. Venus, for instance, is so hot that whatever water it once had has long since evaporated.

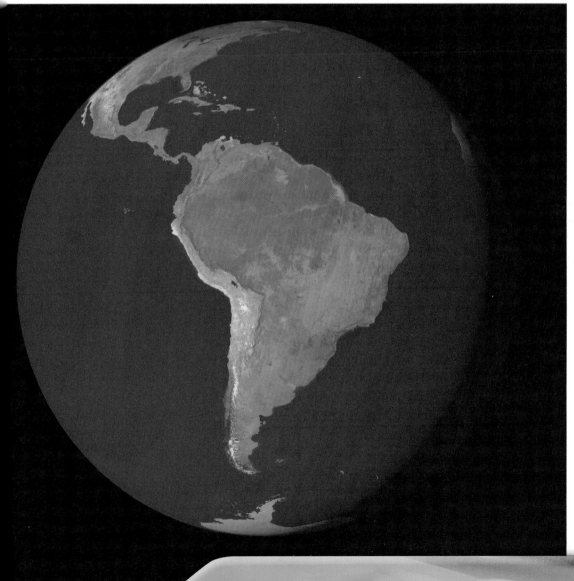

▶ **WHY DO YOU THINK EARTH HAS BEEN CALLED THE "BIG, BLUE MARBLE"?**

PHOTO: NASA

READING SELECTION

EXTENDING YOUR KNOWLEDGE

Mars does harbor frozen water. From space we can see white ice caps—like those on Earth—covering Mars' poles. But Mars' polar ice caps are a bit different from the ones on Earth. They're made of water ice covered by a layer of frozen carbon dioxide, or "dry ice." Scientists are still trying to determine how much water is locked up on Mars in the form of ice. It is possible that Mars was once a warmer, watery planet like Earth. The surface of Mars has many features, such as gullies, that suggest that liquid water was once present. Also, the Mars rovers have found minerals that could only have been formed by evaporation of saline (salty) lakes.

WHAT'S THE FORECAST?

Want to visit one of the gas giants? Bundle up. Because they're far from the Sun, these planets maintain year-round frigid temperatures. Negative 110°C (-166°F) is a typical day on Jupiter. If you're going on a tour of terrestrial planets near the Sun, however, you'll have to be more adaptable. On Mercury, the planet closest to the Sun, the Sun side can heat up to 452°C (846°F) while the side away from the Sun can cool to around -170°C (-274°F). Venus, on the other hand, is always scalding hot, day and night, with temperatures staying around 464°C (867°F). The temperature extremes on Mars are most like what we are accustomed to on Earth, except that those extremes come daily. It might be balmy in the daytime, but it is sure to be freezing at night, with the average temperature a mere -65°C (-85°F): cold enough to freeze antifreeze.

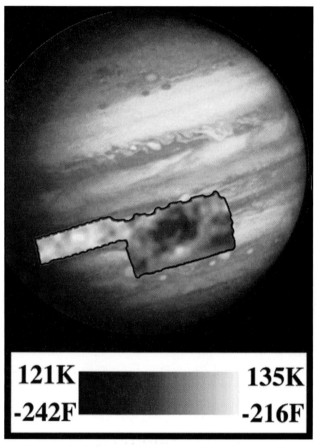

▶ **TEMPERATURE DATA ON JUPITER, COLLECTED BY THE *GALILEO* SPACECRAFT.**

PHOTO: NASA Jet Propulsion Laboratory

ATMOSPHERE CAN BE A HEAVY SUBJECT

Earth's atmosphere allows us to survive on a planet that otherwise would be bombarded by a dangerous level of solar radiation. Composed mostly of nitrogen and oxygen gases, Earth's atmosphere forms a protective barrier that shields against the Sun's DNA-damaging ultraviolet rays. It also protects us from a daily hail of small meteors and other space debris. These pieces of space debris generate so much friction as they fall through the atmosphere that they simply burn up.

Venus also has a robust atmosphere. But the chemical composition of Venus's atmosphere, along with the planet's proximity to the Sun, makes it uninhabitable. Its atmosphere is a thick blanket of carbon dioxide, which produces a severe greenhouse effect. Solar radiation enters, bounces off the planet's surface, and (instead of radiating back into space) remains trapped on Venus, keeping the planet far too hot for human life or even human spacecraft. No unmanned surface mission to Venus has survived on the planet for more than a few hours.

Not all terrestrial planets can boast such robust atmospheres. Mercury has almost no atmosphere at all, and the atmosphere of Mars is quite thin.

The gas giants, as their name suggests, have deep layers of atmosphere surrounding small rocky cores. Jupiter's atmosphere is a human-toxic stew of about half hydrogen, half helium, with traces of other elements, some of which exist as ices. Saturn's bright atmosphere is nearly all hydrogen, with just a pinch of helium. Pale aquamarine Uranus's atmosphere is also hydrogen and helium, with a tiny bit of methane giving it a blue tint. Neptune also has an atmosphere mostly of hydrogen and helium with a trace of methane. The methane gives Neptune its dreamy blue color.

Earth's moon does not have an atmosphere, which is partly why its surface is so heavily dented by falling meteors. It's just too small to have a gravitational pull strong enough to hold onto atmospheric gases. However, many of the gas giants' moons do have atmospheres. Their chemical compositions vary widely. Jupiter's moon, Io, for instance, has a thin atmosphere made largely of sulfur dioxide, while Uranus's moon, Callisto, has a fragile carbon dioxide atmosphere.

It is important to understand that atmospheres change over time. The air you breathe now, for instance, did not exist when Earth was young, four billion years ago. And the atmospheres of the frozen gas giants and their moons have masses of gas that sink, react, and sometimes even escape into space.

READING SELECTION
EXTENDING YOUR KNOWLEDGE

ATMOSPHERE CAN BE A STORMY SUBJECT

Earth's atmosphere is weather-producing, dynamic with tornadoes, typhoons, hurricanes, lightning, and atmospheric currents. Other planets have active atmospheres too. For example, Neptune has a nearly continuous high-pressure storm called the Great White Spot that makes for huge spinning whirlwinds. It boasts the fastest winds in the entire solar system, up to 2500 kilometers (nearly 1600 miles) per hour.

Jupiter nearly rivals it with its Great Red Spot, a cyclone-like storm that's been raging for at least 200 years and is large enough to contain two or three Earths. Despite the thinness of its atmosphere, Mars is no tranquil place either. Winds of more than 200 kilometers (124 miles) per hour regularly whip up dust storms across its surface. These other planets make Earth look like a very comfortable place.

▶ **DUST STORMS HAPPEN ON BOTH MARS (BOTTOM) AND EARTH (TOP).**

PHOTO: NASA Jet Propulsion Laboratory/Malin Space Science Systems

LIFE IS ONE-OF-A-KIND

The conditions of water, atmosphere, and weather make possible Earth's most curious characteristic, the presence of life. Organisms are teeming on land and in water, some visible to us but many of them microscopic.

Scientists do not think that Earth was always this comfortable for life. In its infancy, Earth probably had an atmosphere that was made mostly of hydrogen, helium, and nitrogen. Through volcanic eruptions, carbon dioxide concentrations probably increased slowly until the atmosphere became heavy with the gas. Photosynthetic organisms, tiny creatures that lived in the warm oceans, took in carbon dioxide and released oxygen gas. This gas bound with rocks and other metallic objects, "rusting" them. If you have taken trips to parks with "rusted" or red-banded rocks, you've seen the results.

About two billion years ago, though, the oxygen output exceeded rocks and minerals' ability to absorb it, and oxygen began building up in the atmosphere. This gradually changed the atmosphere of Earth to the oxygen-rich protective one that we know today, which is about 21% oxygen. Along with an atmosphere rich in oxygen came organisms that were able to breathe it, including, eventually, the species we know best: humans. ■

DISCUSSION QUESTIONS

1. In what ways are the planets alike? In what ways are they different?

2. If you were hired to make an advertisement to attract creatures from another planet to visit Earth, what would your ad look like?

THE SOLAR SYSTEM: DESIGNING A SCALE MODEL

INTRODUCTION

About 4.6 billion years ago, our solar system evolved from a vast rotating cloud of cold gas and dust. The cloud separated into chunks of matter and collapsed inward. The Sun formed at the center of this matter as the interior started heating up. Near the Sun, rock and metal came together and eventually became the terrestrial (rocky) planets—Mercury, Venus, Earth, and Mars. The giant, gaseous planets—Jupiter, Saturn, Uranus, and Neptune—formed farther away from the Sun. Asteroids and frozen comets were formed from other loose material.

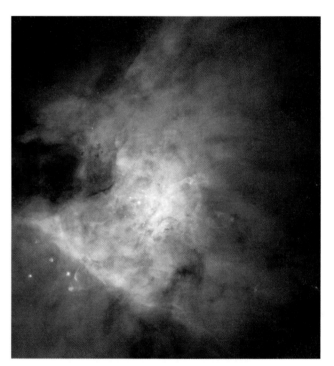

▶ DID OUR SOLAR SYSTEM FORM FROM A GIANT CLOUD OF GAS AND DUST LIKE THIS ONE—THE ORION NEBULA?

PHOTO: NASA Hubble Space Telescope Center

In this lesson, you will begin to investigate the solar system and its planets and asteroids. What do you already know about the solar system? How far apart are the planets? How do the sizes of other planets compare with Earth? You will examine these and other questions as you begin your journey through the solar system.

OBJECTIVES FOR THIS LESSON

- Brainstorm what you know about the order and sizes of the planets and their distances from each other.

- Create a model of the solar system from a set of scaled items.

- Learn how to scale objects in terms of size and distance.

- Use scale models to explore the relative diameters of and distances between the eight planets and the Sun.

- Summarize and organize information about Jupiter.

▶ **MATERIALS FOR LESSON 2**

For you

- 1 copy of Student Sheet 2.1: Our Solar System Model
- 1 copy of Student Sheet 2.2: Using a Scale Factor
- 1 copy of Student Sheet 2.3a: Calculating the Scale Factor
- 1 copy of Student Sheet 2.3b: Calculating Scaled Distance
- 1 copy of Student Sheet 2.3c: Solar System Chart
- 1 calculator

For your group

- 1 large resealable bag labeled "2.1" containing the following:
  - 2 rubber balls
  - 2 ping-pong balls
  - 2 plastic buttons
  - 2 marbles
  - 2 acrylic beads
  - 2 wood barrel beads
  - 2 fishing bobbers
  - 3 split peas
  - 3 pieces of round oat cereal
- 1 strip of adding machine tape
- 1 marker
- 1 small plastic resealable bag labeled "2.3" containing:
  - 2 small round beads
  - 2 peppercorns
  - 1 rubber ball
  - 1 fishing bobber
  - 2 acrylic beads
- 1 metric ruler, 30 cm (12 in.)
- 1 metric measuring tape
- 1 set of 8 Planet Data Cards

GETTING STARTED

1 Record in your science notebook what you already know about the order of, size of, and relative distances between the planets within the solar system. Share your ideas with the class. 🖎

2 Discuss with the class what "scale" means when used with maps. Then define the term "model" and give some examples. Why do you think it is important to build models to scale? Discuss with your class what a scale model of the solar system should look like.

► **WHAT SORTS OF DECISIONS DID THE DESIGNER OF THIS MODEL RAILROAD HAVE TO MAKE? WHAT DECISIONS WILL YOU HAVE TO MAKE IN CONSTRUCTING YOUR MODEL SOLAR SYSTEM?**

PHOTO: D'Arcy Norman/creativecommons.org

DESIGNING A MODEL SOLAR SYSTEM

PROCEDURE

1 Review with your teacher how to convert millimeters to centimeters.

2 Obtain one large resealable bag labeled "2.1." Work with your group to select an object from the bag to represent each planet.

3 Label one end of the strip of adding machine tape "Sun." Position the objects that you have selected along the tape to show the order of, relative sizes of, and relative distances between the planets.

4 List the planets in order on Student Sheet 2.1: Our Solar System Model. Measure the diameter of each planetary model (in centimeters) and record each object's name and diameter. Measure and record the distance (in centimeters) between each model planet and your "Sun."

5 Draw a picture of your solar system model. Label your drawing. Note any similarities in sizes of or distances between your model planets.

6 Share your group's solar system model with the class.

7 Answer the following questions in a class discussion. Answer them in your science notebook as well, if instructed to do so by your teacher. ☞

A. What was the largest planet in your model? What was the smallest planet?

B. Do you observe any patterns in the sizes of the planets in your model? If so, what are they?

C. Do you observe any patterns in the distances between the planets in your model? If so, what are they?

D. Do you think the objects in your model are to scale in terms of size? Why or why not?

E. Do you think the objects in your model are to scale in terms of distance? Why or why not?

8 Clean up by returning all your objects to the resealable bag. Save the adding machine tape for another class, if possible.

INQUIRY 2.2

USING A SCALE FACTOR

PROCEDURE

1 Examine the illustration shown in Figure 2.1. How big would the model of a 10-m-long school bus be if the scale factor that was used to make the model bus is 1 cm = 2 m?

2 Set up your ratio as follows:

$$\frac{X}{10\ m} = \frac{1\ cm}{2\ m}$$

$$X \times 2\ m = 10\ m \times 1\ cm$$

$$X \times 2\ m = 10\ m \times cm$$

$$X = \frac{(10\ \cancel{m} \times cm)}{2\ \cancel{m}}$$

$$X = 5\ cm$$

To calculate the size of the model, divide the actual size of the bus by 2 m, and then multiply the answer by 1 cm. That is, 10 m ÷ 2 m = 5 × 1 cm = 5 cm. The model bus would be 5 cm long.

3 Discuss with your teacher why the ratio "1 cm = 2 m" in Step 1 is considered a "scale factor." How would you define "scale factor"? Record your working definition of this term in your science notebook. ✎

▶ HOW MANY CENTIMETERS LONG WOULD A MODEL OF A 10 m SCHOOL BUS BE IF EVERY CENTIMETER REPRESENTS 2 m?
FIGURE **2.1**

4 Calculate the scaled diameter (SD) of Earth with a scale factor (sf) of 1 cm = 10,000 km (or, 1:10,000). Divide the actual diameter (AD) of Earth (12,742 km) by 10,000 km—the scale factor—and then multiply your answer by 1 cm. Here is what the formula might look like:

$$\text{Scaled Diameter (SD)} : \text{Actual Diameter (AD)} = 1\ cm : \text{scale factor (sf)}$$

$$\frac{\text{Scaled Diameter (SD)}}{\text{Actual Diameter (AD)}} = \frac{1\ cm}{\text{scale factor (sf)}}$$

$$\frac{SD}{AD} = \frac{1\ cm}{sf}$$

$$SD \times sf = AD \times 1\ cm$$

$$SD \times sf = AD \times cm$$

$$SD = \frac{AD \times cm}{sf}$$

or

$$SD = (AD \div sf) \times cm$$

$$SD = AD \div sf$$

5 Review the actual diameters and distances of the planets listed in Table 1 on Student Sheet 2.2: Using a Scale Factor. Calculate the approximate scaled distances and diameters for each planet by using a scale factor of 1 cm = 10,000 km. Record your calculations in the appropriate columns of Table 1. Show all your work.

6 Review your data on Student Sheet 2.2 with the class.

7 Record your answers to the following questions in your science notebook, and then discuss them with your class: ✐

A. How close was your model from Inquiry 2.1 to the scaled model calculated during Inquiry 2.2?

B. Does anything about the data on Student Sheet 2.2 surprise you? Explain.

BUILDING A SCALE MODEL OF THE SOLAR SYSTEM

PROCEDURE

1 Obtain one resealable bag of items labeled "2.3." These items are to scale with the actual diameter measurements of the planets.

2 Which object do you think represents each planet? Position the items so that they represent the sizes of the eight planets (do not consider distances at this point). Record the names of these objects in the third column of Table 1 on Student Sheet 2.3a: Calculating the Scale Factor.

3 Measure the diameter of each of your model planets by placing each object directly onto your ruler or measuring tape (or you can mathematically calculate its diameter from its circumference). Record the diameter of each object in the fourth column of Table 1.

4 Calculate the scale factor for each model planet. Remember to record your units. Use the example on Student Sheet 2.3a to guide you.

5 Calculate the average scale factor by adding all the scale factors together and dividing by eight. Remember to record your units. This average is the approximate scale factor for your entire model solar system.

6 Share your results with the class.

7 Using Student Sheet 2.3b: Calculating Scaled Distance and working with your group, calculate the scaled distance of each planet from the Sun. Refer to the actual distances of the planets from the Sun. Use the scale factor you calculated on Student Sheet 2.3a to determine how far from a model Sun to place each planet. This will help you to create an accurate scale solar system model for both size and distance. Use the example on Student Sheet 2.3b to help you. Note that each answer for scaled distance initially will be in centimeters. However, you will have to convert your answers to either meters or kilometers to make the measurements more meaningful.

8 If possible, go with the class to a long hall or gymnasium, or outdoors to an athletic field. Select one item in the area (for example, a wall or a goal post) to represent the Sun. Then, using your calculations, work with the class to measure the distances for each of the model planets as directed by your teacher. Label each planet with the appropriate Planet Data Card.

REFLECTING
ON WHAT
YOU'VE DONE

1 Answer the following questions in your science notebook, and then discuss them with the class:

A. What observations and comparisons can you make about your model?

B. How does Earth compare with other planets in size and distance from the Sun?

C. How is your model different from the actual solar system?

D. How is your model similar to the actual solar system?

E. What is the relationship between the diameter of the planets and their positions from the Sun? What reasons do you have for your answer?

F. Analyze Table 1 on Student Sheet 2.3b. How do the distances between the first four planets compare with the distances between the other planets? What reasons do you have for your answer?

G. Think back to "Getting Started." How is your final solar system model different from your earlier statements about what a model of the solar system looks like?

2 Obtain one set of Planet Data Cards for your group. Examine the planetary distances and determine what units are used to describe this value. What do you think AU means?

3 Use your calculator to determine how the AU for each planet was calculated. AU is the scale factor most often used when describing planetary distances.

4 Read "Mission: Jupiter" on pages 32–35. Add information about Jupiter to your copy of Student Sheet 2.3c: Solar System Chart. You will use this sheet to record information about the planets as you study this unit.

THE ORRERY: A MODEL OF THE SOLAR SYSTEM

When you build a model of a ship, a rocket, or another object, you usually want it to look as much like the original as possible. When Englishman George Graham built the first mechanical model of the solar system in 1700, he had the same goal. He wanted his model to look like the actual solar system. He duplicated the positions of the Sun and the planets and moons that were known at that time. With Graham's model, people could see for the first time how Earth moved around the Sun. They could also follow the movements of Mercury, Venus, Mars, Jupiter, and Saturn.

Graham's model became known as an "orrery." It was named after Charles Boyce, the 4th Earl of Orrery, who commissioned such a model to be built. The orrery shown here was made in the 1780s. It belongs to the Smithsonian Institution in Washington, D.C. Is there an orrery at a science museum near you?

PHOTO: Smithsonian photo by Eric Long, © 1993, Smithsonian Institution

THE ASTRARIUM: A CLOCK WITHOUT A TOCK

This amazing clock was designed nearly 700 years ago by Giovanni de Dondi. The clock, or astrarium, works like this: One wheel turns once a day and moves another wheel, which makes one complete turn a year. The astrarium tracks the changing positions of Earth, the Moon, the Sun, and the five other planets that were known at the time de Dondi designed the astrarium—Mercury, Venus, Mars, Jupiter, and Saturn. This complex astronomical tool even includes a calendar of all Roman Catholic feasts. De Dondi's invention eventually was dismantled so that its brass could be reused. Luckily, de Dondi left detailed notes about how to construct the astrarium, which were used to make this reproduction.

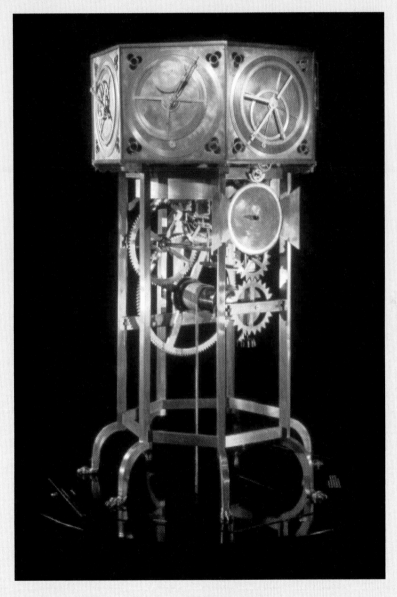

PHOTO: Smithsonian photo by Alfred Harrell, © 1992, Smithsonian Institution

▶ **JUPITER**

PHOTO: NASA Jet Propulsion Laboratory/
University of Arizona

In the early 1600s, astronomer Galileo Galilei looked at cloud-covered Jupiter through one of the world's first telescopes. What he saw amazed everyone—four moons!

Telescopes improved over the years and other moons of Jupiter were found. More than 300 years after Galileo's discovery, spacecraft flew close enough to Jupiter to take detailed pictures. *Pioneer 10* flew by Jupiter first, in 1973. *Pioneer 11* followed a year later. *Voyager 1* and *2*, en route to the outer planets, sped past Jupiter in 1979.

These flyby missions added to our knowledge, but they also left us with many questions. What are Jupiter's rings made of? How fast do the winds on Jupiter blow? And what *is* beneath those clouds?

Many scientists believed that only an orbiter could answer such questions, because an orbiter can observe a planet for a long time. The *Viking*

orbiter had revealed much about Mars. And the *Pioneer* orbiter answered many questions about Venus. But how could an orbiter "see" beneath Jupiter's clouds? A probe would have to do that!

Plans got underway to send a spacecraft with an orbiter *and* a probe to Jupiter. The mission would be named after the Italian scientist who first spotted Jupiter's four large moons—Galileo.

The *Galileo* spacecraft was launched in 1989. Six years later, as it approached Jupiter, the probe and orbiter separated. In December 1995, the probe finally plunged into Jupiter's clouds. The orbiter, meanwhile, was starting on its path around the largest planet in the solar system.

GALILEO PROBE FINDINGS

Descending through Jupiter's atmosphere, frictional forces on the surface of the probe's heat shield raised the heat shield's surface

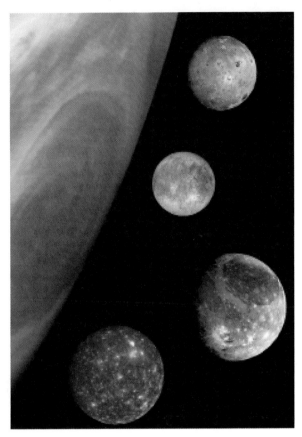

▶ JUPITER'S FOUR LARGEST MOONS: IO, EUROPA, GANYMEDE, AND CALLISTO (TOP TO BOTTOM). JUPITER'S GREAT RED SPOT IS SHOWN IN THE BACKGROUND.

PHOTO: NASA Jet Propulsion Laboratory/DLR (German Aerospace Center)

▶ THE *GALILEO* PROBE'S DESCENT THROUGH JUPITER'S ATMOSPHERE

PHOTO: NASA

temperatures to levels twice as hot as those of the Sun's surface! Still, it sent data for nearly 58 minutes before the heat and pressure destroyed it.

What can a space probe discover in less than an hour? As it turns out—lots! To the surprise of scientists, the probe showed there were no clouds in the lower part of Jupiter's atmosphere. The air below the clouds also was much drier than scientists expected. Scientists think the probe may have descended into a part of Jupiter's atmosphere that was unusually dry. Yet the wind speed—540 kilometers (336 miles) per hour—was the same both in the clouds and below them.

These winds puzzled scientists.

On Earth, the Sun's heat helps make winds. But Jupiter receives only about 1/25 as much sunlight as Earth. So what is the source of Jupiter's winds? Some scientists believe they are powered by heat escaping from deep inside the planet.

The probe journeyed 600 kilometers (373 miles) into Jupiter's atmosphere. As expected, it hit no solid object or surface along the way. Jupiter is, after all, a gaseous giant. Its solid surface lies below tens of thousands of kilometers of atmosphere.

READING SELECTION
EXTENDING YOUR KNOWLEDGE

▶ THE RING SYSTEM OF JUPITER, SHOWN HERE, WAS IMAGED BY THE *GALILEO* SPACECRAFT ON NOVEMBER 9, 1996.

PHOTO: NASA Jet Propulsion Laboratory

GALILEO ORBITER FINDINGS

In the late 1970s, the *Voyager* spacecraft discovered two, possibly three, rings around Jupiter. In 1995, the *Galileo* orbiter confirmed the presence of a third thin ring but also found a fourth ring inside it! According to the data transmitted by *Galileo*, all the rings consist of small grains of dust. It seems that meteoroid impacts blasted the grains off the surface of the four innermost moons.

Galileo also revealed that Jupiter is home to many more huge storms than once thought. The largest is the Great Red Spot. This raging storm is three times the size of Earth!

While Jupiter has more storms than expected, it has less lightning than scientists predicted. Lightning occurs on Jupiter only about one-tenth as often as on Earth. On our planet, there is an average of about 100 lightning flashes every second.

While orbiting Jupiter, *Galileo* flew near several of Jupiter's moons. Scientists knew that Io, the innermost of Jupiter's four major moons, has active volcanoes. The *Voyager* spacecraft discovered several of them in 1979. But the *Galileo* orbiter showed that hundreds of volcanoes cover Io. Many spew lava from deep below the moon's surface.

Jupiter's moon Europa seems to have an ocean beneath its cracked icy surface. In this ocean, gigantic blocks of ice the size of cities appear to have broken off and drifted apart. Callisto, the outermost of Jupiter's four major moons, also may have an ocean below its cratered surface.

Ganymede is the largest moon in the entire solar system. The data transmitted by the

▶ THE GREAT RED SPOT ON JUPITER

PHOTO: NASA Jet Propulsion Laboratory

Galileo orbiter show that it is the only moon known to have a magnetic field.

The orbiter ended its mission with a dive into Jupiter's deadly atmosphere—a dramatic ending for one mind-expanding trip! ■

DISCUSSION QUESTIONS

1. Why is Jupiter called a gaseous planet when it has some solid surface?

2. What is the value of sending orbiters and probes to other planets?

PLANETARY FACTS

JUPITER

JUPITER: QUICK FACTS*

▶ **Diameter**
139,822 km

▶ **Average distance from the Sun**
778,340,821 km

▶ **Mass**
$189,813 \times 10^{22}$ kg

▶ **Surface gravity (Earth = 1)**
2.53

▶ **Average temperature**
-110°C

▶ **Length of sidereal day**
9.92 hours

▶ **Length of year**
11.86 Earth years

▶ **Number of observed moons**
62

COMPOSITION

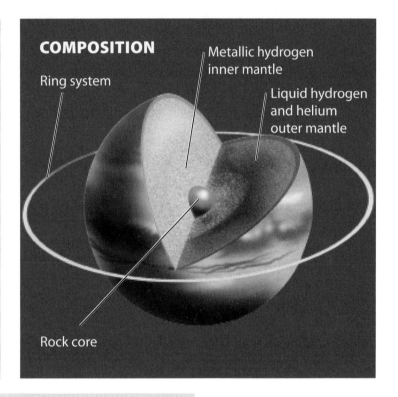

Ring system

Metallic hydrogen inner mantle

Liquid hydrogen and helium outer mantle

Rock core

RELATIVE SIZE

DID YOU KNOW?

▶ Jupiter (like the other gaseous planets) has a ring system. Although Jupiter's ring system is essentially transparent, the ring particles are visible when there is light behind them.

▶ Jupiter's swirling cloud patterns are caused by its rapid rotation.

ATMOSPHERE

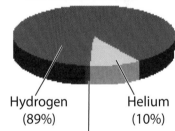

Hydrogen (89%)

Helium (10%)

Methane, ammonia, and water (traces)

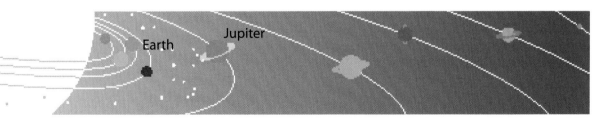

Earth

Jupiter

* Source: Data from NASA as of 2011

SURFACE GRAVITY

▶ ON THE MOON, THIS ASTRONAUT WEIGHS ONE-SIXTH OF HIS WEIGHT ON EARTH. THIS IS BECAUSE THE MOON'S SURFACE GRAVITY IS NOT AS STRONG AS EARTH'S.

PHOTO: NASA Johnson Space Center

INTRODUCTION

How much do you weigh? How massive are you? Suppose you could travel to other planets. What would happen to your weight and mass? In this lesson you will explore these questions. You will simulate what it is like to pick up objects on different planets. Then, you will read about the difference between mass and weight. The lesson ends as you continue the "Mission" series of reading selections to learn more about the missions to Venus.

OBJECTIVES FOR THIS LESSON

Use a model to compare the weight of a can of soda on different planets.

Analyze the relationship between an object's weight on each planet and the planet's mass and diameter.

Describe how mass and weight (force of gravity) are related.

Summarize and organize information about Venus.

MATERIALS FOR LESSON 3

For you

1	copy of Student Sheet 3.1: How Much Would a Can of Soda Weigh?
1	working copy of Student Sheet 2.3c: Solar System Chart

For your group

8	prepared cans
1	set of Planet Data Cards

GETTING STARTED

1 What do you know about gravity as a characteristic of a planet's surface? Record your ideas in your science notebook. 🖉

2 Share what you have written with the class and discuss your ideas about the following questions:

A. What is gravity?

B. How are gravity and weight related?

C. How can you measure gravity?

▸ **WHAT EVIDENCE OF GRAVITY DO YOU SEE IN THIS PHOTO?**

PHOTO: by Jurvetson (flickr)

ANALYZING WEIGHT ON EACH PLANET

PROCEDURE

1 Examine the prepared cans at your assigned station. Every can represents the same full can of soda but on different planets. Pick up each can and observe how heavy that can is on each planet. On which planet does the can weigh the least? On which planet does the can weigh the most? Why do you think this is? Discuss your ideas with your group.

2 Rank the weight of each can on Student Sheet 3.1: How Much Would a Can of Soda Weigh? by recording a "1" under the name of the planet with the lightest can. Increase the numbers until you get to the heaviest can. If any cans seem to weigh the same, use the same number to rank them.

3 Turn over the following three Planet Data Cards: the planet you ranked "1" (the lightest can); the planet you ranked with the highest number (the heaviest can); and a planet with a number somewhere in between. Examine the data printed on the back of these three cards. What characteristics about each planet could explain why the weight of the can is different on each planet? Discuss your ideas with your group. Record your ideas below Table 1 on Student Sheet 3.1.

REFLECTING
ON WHAT
YOU'VE DONE

1 Share your rankings from Student Sheet 3.1 with the class.

2 How do the characteristics of a planet affect its ability to pull on an object—giving the object its weight? Share your ideas with the class, using your explanation on Student Sheet 3.1 as your guide.

3 Record in your science notebook what you know about the relationship between mass and weight. Then share your ideas with the class.

4 Read "Mass and Weight: What's the Difference?" on pages 40–43. Then record a working definition for each of these two terms in your science notebook.

5 If weight is the measure of the force of gravity pulling on an object, which planet has a greater force of gravity pulling on objects at its surface—Mercury or Jupiter? Explain your answer.

6 What two factors affect the gravity at a planet's surface?

7 If Mars has more mass than Mercury, why is the force of gravity on the surface of Mars nearly the same as the force of gravity on the surface of Mercury?

8 With your class, return to the Question D (If you could travel to another planet, what would happen to your weight?) folder from Lesson 1. Is there anything you would now change or add? Discuss your ideas with the class.

9 Read "Mission: Venus" on pages 44–49. Add any information about Venus to your working copy of Student Sheet 2.3c: Solar System Chart.

MASS AND WEIGHT

What's the Difference?

Many people think that there is no difference between the terms "weight" and "mass." But there is! Mass is related to the amount of matter (or "stuff") in an object, regardless of how much space the object takes up. As long as you do not add or take away any matter from an object, its mass stays the same, even at different locations. If you take an object to the Moon or Mars, it will have the same mass that it had on Earth.

But what about its weight? Would an object weigh the same on the Moon or Mars as it does on Earth? As you found out in your investigation, the answer is "no." The weight of an object changes from planet to planet. Weight even changes from one place on a planet (such as a mountaintop, where you might weigh less) to another (such as the bottom of a deep valley, where you might weigh more).

MEASURING MASS AND WEIGHT

Weight is a measure of the force of gravity on an object. (A force is a push or pull on or by an object.) We use a spring scale to measure the strength of the gravitational pull on the mass of an object. Objects with the same mass have the same weight. An object with more mass has a stronger gravitational force pulling on it than an object with less mass. The spring scale is pulled down farther, showing that the object weighs more. In the metric system, weight is measured in newtons.

Mass is measured in grams and kilograms in the metric system. To find the mass of an object, we use a balance. When equal amounts of matter are placed on opposite sides of a beam balance, for example, the pull of gravity is the same on both sides, and the beam balances. If the mass of an object were measured with a balance on Earth, and then with the same balance on the Moon, the results would be identical. The amount of "stuff" in the object hasn't changed.

WEIGHT IS MEASURED USING A SPRING SCALE. MASS IS MEASURED USING A BALANCE.

MASS AND WEIGHT ON THE PLANETS

An object with mass attracts any other object with mass. The strength of that attraction depends on the mass of each object and the distance between those objects. This gravitational pull is very small between objects of ordinary size and therefore is hard to measure. The pull between an object with a large amount of mass, such as Earth, and another object, such as a person on the planet's surface, can easily be measured.

Weight on a planet's surface is a measure of the pull of gravity between an object and the planet on which it is located. This force of gravity on an object on a planet's surface depends on the object's mass and the mass of the planet. If the mass of that object is doubled, gravity pulls on it twice as hard. If the mass of the planet is doubled, gravity pulls on the object twice as hard.

READING SELECTION
EXTENDING YOUR KNOWLEDGE

WHAT WOULD YOU WEIGH ON JUPITER?

Jupiter has 318 times more mass than Earth, so you might assume that you would weigh 318 times more on Jupiter than you weigh on Earth. This would be true if Jupiter were the same size as Earth, but the diameter of Jupiter is more than 10 times the diameter of Earth. This means that if you stood on Jupiter, you would be farther from the planet's center than you would be if you stood on Earth. This reduces Jupiter's gravitational pull on you to only about 2.53 times (and not 318 times) your weight on Earth.

The number 2.53 is referred to as Jupiter's "gravity factor." The gravity factor is the ratio of each planet's gravity to that on Earth. Earth's gravity factor is 1 and Jupiter's gravity factor is 2.53. By multiplying your Earth weight by a planet's gravity factor, you can determine your weight on that planet. Use the table to find out how much you would weigh on each of the eight planets.

TABLE 1 MASS, RADIUS, AND SURFACE GRAVITY OF EACH PLANET

PLANET	MASS (10^{22} kg)	RADIUS (km)	SURFACE GRAVITY FACTOR (EARTH = 1)
MERCURY	33	2440	0.38
VENUS	487	6052	0.90
EARTH	597	6371	1.00
MARS	64	3390	0.38
JUPITER	189,813	69,911	2.53
SATURN	56,832	58,232	1.06
URANUS	8681	25,362	0.90
NEPTUNE	10,241	24,622	1.14

Source: http://solarsystem.nasa.gov

The force of gravity also depends on the distance of an object from the center of the planet to its surface. This distance is called the radius of the planet. The farther an object is from the planet's center, the weaker the pull between the planet and the object. This force gets weaker quite rapidly, but there is a pattern to it. If you double the radius of the planet, the weight of the object will be one-fourth as much. If you triple the radius of the planet, the weight will be only one-ninth as much. The force of gravity drops off with the square of the distance between the center of the planet and the object.

Each planet in our solar system has a different mass and a different size. This means that the weight of the same object on the surface of each planet will be different. For example, you would weigh less on the Moon than you do on Earth because although the Moon is smaller than Earth, it also has less mass than Earth—and Earth's mass wins out. This means that the Moon exerts less gravitational force at its surface than Earth. Any given object will have the same mass on Earth and on the Moon, but that object's weight on the Moon will be only about 16 percent (one-sixth) of the weight as measured on Earth. ■

▶ THE FARTHER THE OBJECT IS FROM THE CENTER OF THE PLANET, THE WEAKER THE PULL BETWEEN THE PLANET AND THE OBJECT. THIS MEANS THAT YOUR WEIGHT ON SATURN WOULD BE CLOSE TO YOUR WEIGHT ON VENUS, EVEN THOUGH SATURN IS MORE MASSIVE THAN VENUS.

? DISCUSSION QUESTIONS

1. Where would you have more mass: on Earth or on Neptune? Explain your answer.

2. What factors would you have to take into account to determine your weight on Mars?

mission: Venus

Sky gazers and astronomers have known about Venus for thousands of years. That's because it's the brightest object in our sky, except for the Sun and Moon. By looking through powerful telescopes over the last 100 years, we've learned much about our neighboring planet. Astronomers have discovered, for example, that thick clouds surround Venus, explaining the brightness of the "evening star," as it's still sometimes known. The thick Venusian blanket of clouds reflects nearly 70 percent of the nearby Sun's light. If Earth is a big blue marble, Venus is a dazzling yellowish-white one.

Still, scientists have wondered as they gazed at Earth's "sister planet." What is the weather like on Venus? Is there water? Wind? Life?

THE *VENERA* PROBES

The earliest Venus explorers were the Soviets, sending a battery of thirteen missions to Earth's closest neighboring planet from the 1960s through the 1980s. These were the Venera missions. Three of the spacecraft merely flew by Venus, recording what they could from space. Ten entered the planet's atmosphere and dropped to the surface, exploring the soil and sending images of the planet back to scientists in Moscow. Conditions on Venus are so harsh—extremely hot, with a corrosive atmosphere—that none of the landers lasted more than about an hour, a remarkably short time given the tremendous effort and expense of sending a capsule to Venus. In those hours, though, the landers successfully relayed sharp images of the rocky planet, drilled into the surface, and discovered what those rocks were made of. They tested the atmosphere and found what was in the Venusian air. This gave scientists around the world a front-row view of our sister closer to the Sun, where no human explorer could survive.

The Venera probes racked up many space-exploration firsts, including, in 1970, the first soft landing on another planet's surface. The

▶ THIS IMAGE OF VENUS IS MADE UP OF RADAR
IMAGES FROM THE SPACECRAFT *MAGELLAN*.

PHOTO: NASA Jet Propulsion Laboratory

READING SELECTION

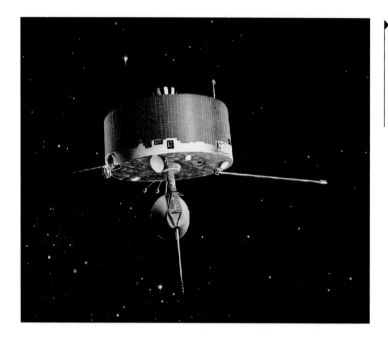

▶ THE *PIONEER* VENUS ORBITER (*PIONEER 12*) SPACECRAFT IS SHOWN IN ITS NORMAL FLIGHT ATTITUDE (UPSIDE DOWN)

PHOTO: NASA Ames Research Center

most important data they sent back to Earth, however, included evidence that Venus was a volcanic planet. Its surface seemed to be made of ridged lava plains, pulverized by meteors to a rocky soil in some places. Venera probes measured the wind speeds in the atmosphere and found that the clouds—tens of kilometers thick—rotated tremendously fast, at speeds of up to 100 meters (328 feet) per second.

The Soviet successes were a real challenge to American scientific pride. In 1978, the United States launched two spacecraft to Venus in search of its own answers. These two Pioneer mission spacecraft were followed 11 years later with the launch of *Magellan*.

PIONEER 12—THE ORBITER

The first of the two Pioneer missions, the *Pioneer 12* Venus Orbiter, took off in May 1978. It was in orbit around Venus by December—and it stayed there for 14 years. At times, the orbiter was as close as 150 kilometers (93 miles) to Venus's surface.

During its long stay in orbit, *Pioneer 12* measured, mapped, and photographed Venus's yellow cloud cover. Information from *Pioneer 12* showed that the upper layers of clouds contained droplets of sulfuric acid, a poisonous liquid that can destroy metals.

Pioneer 12 also proved what many scientists suspected—that a powerful "greenhouse effect" exists on Venus. This means that Venus's atmosphere, which is made up mostly of carbon dioxide, traps heat from the Sun. The heat can raise the surface temperature of Venus to about 475°C (887°F)—hot enough to melt lead! Venus may have constant lightning flashes and rain, but the intense heat means that the rain evaporates before it reaches the ground. If Venus's surface once had water, it has all boiled away and been lost to space. Today, the planet is very dry.

On October 8, 1992, *Pioneer 12* ran out of fuel. As it fell through the clouds of Venus, the orbiter burned up. It is gone forever, but the data and photos it radioed to Earth remain.

PIONEER 13—THE MULTIPROBE

The second Pioneer spacecraft, *Pioneer 13* Venus Multiprobe, was launched in August 1978. *Pioneer 13* was designed to enter Venus's atmosphere rather than to orbit the planet. It was equipped with one large probe and three small probes. Probes are packed with instruments to take observations and measurements such as atmospheric content, turbulence, temperature, particle size, and radiation. The large probe was 1 meter long and weighed 315 kilograms (694 pounds). The small probes were 0.8 meters (2.6 feet) in diameter and weighed 75 kilograms

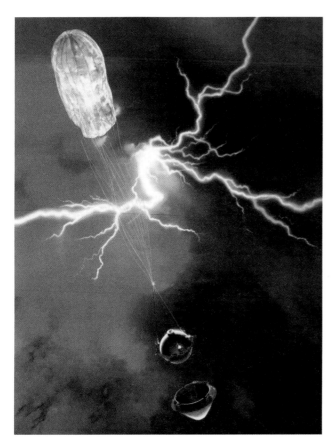

▶ ARTIST'S CONCEPT OF A PIONEER VENUS PROBE ENTERING VENUS'S ATMOSPHERE

PHOTO: NASA National Space Science Data Center

(165 pounds) each. On December 9, 1978, each of the probes descended from the spacecraft for about 55 minutes before hitting the planet's surface.

These probes were not designed for a soft landing, as the Venera spacecraft had been. Engineers who designed the Pioneer probes knew they would not withstand the impact of landing on the planet's surface. It was the information the probes collected on the way down to the surface that would be most important. However, one small probe did survive its impact with Venus's surface and continued to send data from the surface for 67 minutes before the heat and pressure ended its ability to transmit.

As they made their way through the atmosphere, all four probes captured a great deal of information. Data showed, for example, that winds in the upper layers of Venus's atmosphere move very fast. They circle the planet every four days! Winds near the surface, though, are strong enough only to move sand grains and dust particles.

Sensors on the probes also recorded atmospheric pressure on Venus. Venus's thick atmosphere pushes on things with 90 times the force of Earth's atmosphere. This kind of crushing pressure exists on Earth only about one kilometer down in the ocean.

MAGELLAN

Named after the 16th-century Portuguese explorer who was the first to sail around the world, the *Magellan* spacecraft was launched on May 4, 1989. It arrived at Venus on August 10, 1990. For five years, *Magellan's* large main antenna bounced radar signals off the surface of Venus, checking to see how fast the signals came back. This gave us information about where the peaks and valleys lay on Venus's surface. The longer the signal took to bounce back, the deeper into the surface it had gone. All radar data were transmitted back to Earth, where a computer used the data to create images of the planet's surface.

READING SELECTION

EXTENDING YOUR KNOWLEDGE

▶ **VOLCANISM ON VENUS**

PHOTO: NASA Jet Propulsion Laboratory—Caltech/ European Space Agency

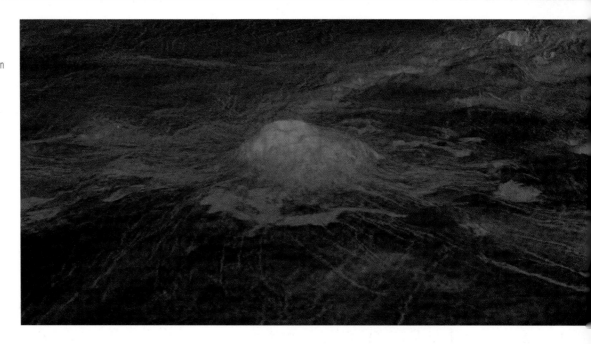

VOLCANISM ON VENUS

Magellan confirmed that Venus is a planet of volcanoes. Thousands of them dot the planet— many small domes and some huge mountains. Smooth plains formed by flowing lava separate many of the volcanoes. Lava has also carved channels—some as long as the width of the continental United States—on the surface of Venus.

Magellan also showed that the surface of Venus has been pulled apart in some places, possibly by the upwelling of magma from Venus's mantle. One such rift, or area of weakened crust, tore a crater in two. Volcanoes often occur along these rift zones.

VENUS *EXPRESS*

In 2005, the European Space Agency sent a probe on a mission to Venus. The probe was called Venus *Express*. It began sending data back in 2006. Venus *Express* extensively explored the atmosphere of Venus and the winds at different levels of the atmosphere. The instruments on Venus *Express* allowed scientists to get a 3-D view of the wind

speeds and directions on Venus. The probe confirmed that the winds on Venus are very fast and that wind speed varies at different levels in the atmosphere. It also allowed scientists to investigate the nature of a huge vortex in the atmosphere at the poles of Venus. Studying wind patterns on Venus allows scientists to learn about the dynamics in the atmosphere of this hot planet. ∎

DISCUSSION QUESTIONS

1. What about Venus makes it similar to Earth? What makes it different?

2. What are some of the challenges in collecting data from planets like Venus?

PLANETARY FACTS

VENUS

VENUS: QUICK FACTS*

- **Diameter**
  12,104 km
- **Average distance from the Sun**
  108,209,475 km
- **Mass**
  487 x 10^{22} kg
- **Surface gravity (Earth = 1)**
  0.90
- **Average temperature**
  464°C
- **Length of sidereal day**
  243.02 Earth days (retrograde)
- **Length of year**
  224.70 Earth days
- **Number of observed moons**
  0

COMPOSITION

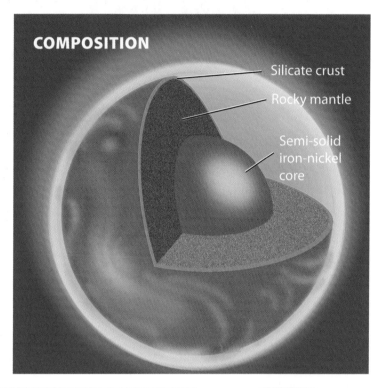

Silicate crust

Rocky mantle

Semi-solid iron-nickel core

RELATIVE SIZE

DID YOU KNOW?

- Venus rotates from east to west—the opposite direction of Earth.
- Venus's gas clouds reflect the Sun's rays so well that Venus shines more brightly than any other planet.

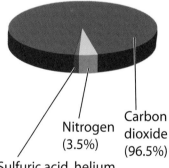

Nitrogen (3.5%)

Carbon dioxide (96.5%)

Sulfuric acid, helium, argon, and others (minor amounts)

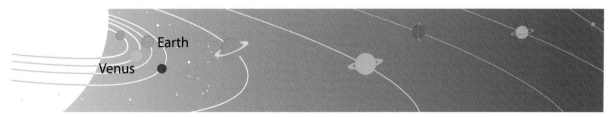

Earth

Venus

* Source: Data from NASA as of 2011

EXPLORATION ACTIVITY: SOLAR SYSTEM EXPLORATIONS

INTRODUCTION

For much of human history, our knowledge about space has been limited. Information that we obtained about space came either from our own observations or from the observations made by instruments set up on Earth. Only since the 1960s have we been able to leave Earth and experience space firsthand. Since then, space exploration has literally opened up new worlds.

In this lesson you will begin a research project (called the Exploration Activity) to learn more about the planets and other objects in the solar system. First, you will read about the history of space exploration and what we have learned from those missions.

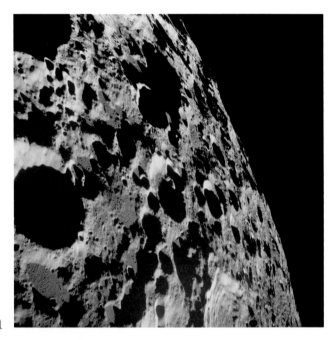

IMAGE OF THE MOON'S CRATERS SHOT IN 1969 BY *APOLLO 11*, THE FIRST MISSION TO LAND PEOPLE ON THE MOON. HOW DO YOU THINK THE *APOLLO 11* ASTRONAUTS FELT AS THEY APPROACHED THE MOON'S SURFACE?

PHOTO: NASA Johnson Space Center

You will also read about some risks associated with space exploration and how unmanned probes are used to explore the solar system. Then, after you review the guidelines for the Exploration Activity, you will use different resources over the next few weeks to learn more about a particular planet or other solar system object and probes that have observed it. You will record your research in a solar system travel brochure. Then, with your team members, you will use the information to design a space mission to explore that object in the future. In a few weeks you will present your brochure to the class and your team will share its future mission design.

OBJECTIVES FOR THIS LESSON

Read about the history of the space program.

Discuss how science and technology contribute to the advancement of planetary studies.

Research information about a planet or other solar system object.

Design a travel brochure about the solar system object.

Design a space mission for exploring the solar system object in the future.

Summarize and organize information about Earth.

▶ **MATERIALS FOR LESSON 4**

For you

1   working copy of Student Sheet 2.3c: Solar System Chart

1   copy of Inquiry Master 4.1a: Exploration Activity Scoring Rubrics

1   copy of Student Sheet 4.1a: Exploration Activity Timeline

1   copy of Student Sheet 4.1b: Solar System Brochure Outline

1   sheet of white or light-colored paper, 8½ × 11 in

Markers
Colored paper
Art supplies

GETTING STARTED

1 Read "The History of Space Exploration" on pages 58-65.

2 With your group, select one space mission (such as Gemini) or a space probe (such as *Sojourner*) from the reading selection. Report to the class what you learned or already know about that mission or probe.

3 Discuss the following with your class:

A. How has space exploration changed since the 1960s?

B. What have we learned about the solar system from the space program?

C. What do you think will happen to space exploration in the future?

PART 1
BEGINNING THE EXPLORATION ACTIVITY
PROCEDURE

1 Your teacher will show you a sample travel brochure. Discuss its features.

2 Obtain one white or light-colored sheet of paper and practice folding it in thirds to make a brochure. See Figure 4.1. You can use this paper for the rough draft of your solar system travel brochure. You are free to use paper or material of your own choosing for your final brochure.

3 Your teacher will give you a copy of Student Sheet 4.1a: Exploration Activity Timeline. Discuss the due dates with your teacher, then record them on your student sheet.

4 Discuss with your teacher how your project will be assessed.

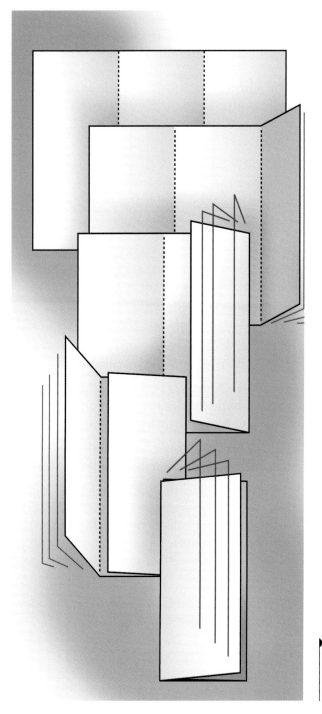

5 Select a specific planet or other solar system object for your Exploration Activity. Record the name of that solar system object at the top of Student Sheet 4.1a. Tell the class which solar system object you have selected.

NOTE Remember that Earth is a planet and is observed from space just as other planets are.

6 On Student Sheet 4.1a, record the names of one to three other students who will be researching your solar system object. You will work with these team member(s) to design a mission to explore the solar system object in the future. You and your team will share your future mission design in a few weeks.

▶ **FOLD THE PAPER TO CREATE A BROCHURE.**
FIGURE **4.1**

PART 2
CONDUCTING YOUR RESEARCH

PROCEDURE

1. Review Student Sheet 4.1b: Solar System Brochure Outline and Student Sheet 2.3c. Use these sheets as a way to organize your research. Record the due date for your outline at the top of Student Sheet 4.1b.

2. Go to the library or the computer lab with your class. Conduct research to find out more about your solar system object. You should use a variety of resources, including those in the classroom, in your school library and other libraries, in newspapers, or on the Internet. You also may conduct personal interviews or use e-mail to ask questions of experts. The "Mission" reading series in your Student Guide is another excellent resource. Your teacher will tell you how many Internet sources, books, magazines, and newspaper articles you need to use in your research.

3. Complete Student Sheet 4.1b and fill in the row for your solar system object on Student Sheet 2.3c as you conduct your research. Find out as much as you can about your solar system object and about space probes and missions that have provided this information. Use the "Mission" series to help you. Note that you may not be able to find answers to all of the questions on Student Sheet 4.1b. With your teacher's permission, you may substitute other information about your object. You also might want to find pictures of your object to put in your finished brochure.

You should include the following information in your brochure (these items are also listed on Student Sheet 4.1b):

HISTORY: The object's name, who named it, and what the name means

DISCOVERY: When the object was discovered and who discovered it

OBJECT STRUCTURE: The interior of the object (what it looks like inside) and its surface features (if it is terrestrial)

ATMOSPHERE: The composition and conditions of the object's atmosphere (for example, does it have storms?), if these are known; if there is no atmosphere, say so

MOTION: The object's orbit (how long it takes to get around the Sun) and rotation (how long it takes to turn one time on its axis)

MISSIONS: A description of probes or missions to the object, including travel times and dates

DATA: The information you entered on Student Sheet 2.3c for your object. Include the object's diameter, average distance from the Sun, mass, surface gravity, average temperature, length of sidereal day, length of year, and number of observed moons.

OTHER: Other interesting information (include pictures and drawings)

4. Create your bibliography. As you conduct your research, remember to use your school's guidelines for giving credit to your sources, including any images you may be including in your brochure.

PART 3
PRESENTING WHAT YOU'VE LEARNED

PROCEDURE

1 Make your solar system brochure. Prepare your travel brochure using the information that you outlined on Student Sheets 4.1b and 2.3c. Include both pictures and text. Try to be original. See Figure 4.2 for examples of completed brochures.

▶ **SAMPLE BROCHURES**
FIGURE **4.2**

PHOTO: Jeff McAdams, Photographer, Courtesy of Carolina Biological Supply Company

Exploration Activity Part 3 continued

 2 Work as a team with other students in your class who researched the same solar system object to design a way to explore it in the future. You will consider all the knowledge you gained about your solar system object to determine if there are any travel constraints and to design your exploration to overcome these constraints. Consider questions such as the following:

- How does the distance to the object determine the type of spacecraft you must use to probe it firsthand?

- Will the probe land on the object's surface or observe it from space?

- Can your probe deal with the object's gravitational forces?

- How will the object's surface conditions and atmospheric composition affect the type of observations you can make?

▶ **SHARE YOUR TRAVEL BROCHURE WITH YOUR CLASS.**
FIGURE **4.3**

PHOTO: Jeff McAdams, Photographer, Courtesy of
Carolina Biological Supply Company

3. Present your brochure and future mission design to the class. (See Figure 4.3.) Be prepared to debate and discuss any issues or topics that arise.

REFLECTING
ON WHAT YOU'VE DONE

1 With your group, compare the eight planets in each category listed on Student Sheet 2.3c. Look for patterns and exceptions to the patterns. (For example, does the distance of a planet from the Sun relate to its mass? Does the mass of a planet determine how many moons it can have? If a pattern in mass and distance exists, is there a planet that doesn't follow this pattern?) Decide with your group how to record the patterns you discover.

2 Analyze any patterns or exceptions to the patterns by answering the following in your science notebook, then discuss them as a class:

A. Which categories listed on Student Sheet 2.3c seem to be related? Explain why you think they are related.

B. Which categories seem to stand alone, with little or no relation to the others?

C. Which planet breaks a pattern and in which category?

D. Look at the "Mission" reading selections in Lessons 2–7. What patterns do you observe in the planets' atmospheres? Which categories seem to be related to whether a planet has an atmosphere? Explain your thinking.

3 Read "Climate's Link to Life" on pages 73–77. Answer the questions at the end of the reading selection in your science notebook, then discuss your answers with the class.

THE HISTORY OF SPACE EXPLORATION

▶ ASTRONAUT BUZZ ALDRIN
DEPLOYING A SCIENTIFIC
EXPERIMENT PACKAGE DURING THE
FIRST MANNED SPACE FLIGHT TO
THE MOON IN 1969

PHOTO: NASA Marshall Space Flight Center

The story of space exploration begins in the early 20th century. Before space flight was even physically possible, humans imagined being able to explore space. Human space flight took a giant leap toward reality during World War II, when Germany began to manufacture V2 rockets. The United States captured many of these rockets after the war and used them for testing and further development. In the late 1950s, scientists developed winged rocket planes based on their studies of those earlier V2 rockets. In 1957, the former Soviet Union sent the first satellite, *Sputnik 1*, into orbit. In 1961, the Soviet Union's air force pilot and cosmonaut Yuri Gagarin became the first human to orbit Earth.

YURI GAGARIN, A SOVIET
AIR FORCE PILOT AND
COSMONAUT, WAS THE FIRST
HUMAN TO ORBIT EARTH.
HIS TOTAL FLIGHT TIME WAS
1 HOUR AND 48 MINUTES.

PHOTO: NASA

READING SELECTION
EXTENDING YOUR KNOWLEDGE

MERCURY–ATLAS ROCKET
STANDS ON A LAUNCH PAD AT
CAPE CANAVERAL AWAITING A
MISSION IN 1963.

PHOTO: NASA Kennedy Space Center

PROJECT MERCURY

The National Aeronautics and Space Administration (NASA) was formed on October 1, 1958. Six days later, NASA announced that its goal was to place an American in orbit. Building on the groundwork laid by a group working at Langley Air Force Base, NASA began its first high-profile program—Project Mercury. At that time, no human had ever gone into space. No one knew how astronauts would react to weightlessness, the high accelerations of launches and landings, radiation from the Sun, or the psychological stresses of being in space. So NASA first sent chimpanzees, monkeys, and human-like dummies into space. Scientists tested the tracking and recovery systems of various spacecraft. NASA announced that seven astronauts, all military men, would train intensely to prepare for space travel.

Project Mercury flew 13 flights—six of which were piloted. On May 6, 1961, Alan Shephard rocketed 185 kilometers (115 miles) above Earth in a flawless flight to become America's first man in space, 23 days after Yuri Gagarin's flight. After Shephard's flight, President John F. Kennedy announced NASA's next space program goal: to land a man on the Moon by the end of the decade.

Project Mercury proved that astronauts could control their spacecraft while in space, and that a human could remain in space for more than a day.

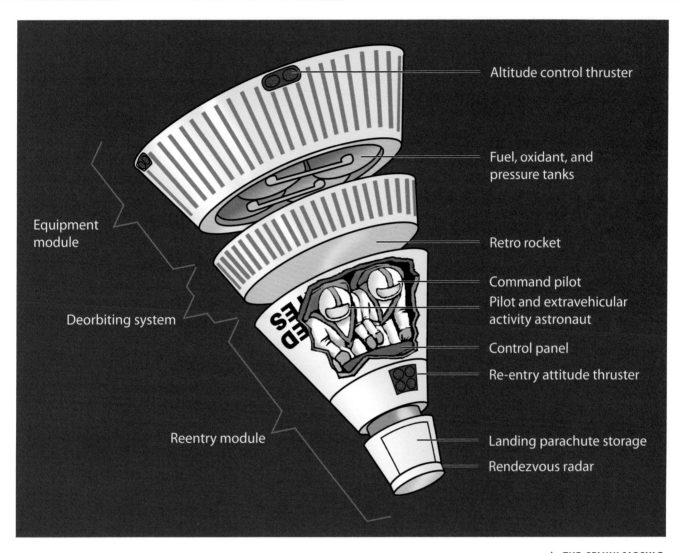

Altitude control thruster

Fuel, oxidant, and pressure tanks

Equipment module

Retro rocket

Command pilot

Pilot and extravehicular activity astronaut

Deorbiting system

Control panel

Re-entry attitude thruster

Reentry module

Landing parachute storage

Rendezvous radar

▶ THE *GEMINI* CAPSULE

PROJECT GEMINI

Project Gemini built on Project Mercury's successes and used spacecraft built for two astronauts. It also developed the United States' ability to maintain crews on extended space missions and to maneuver and rendezvous, all of which were critical to the later Apollo missions. Gemini astronauts also conducted several scientific experiments and performed the first U.S. "space walks."

A total of 12 Gemini missions flew, two of them unpiloted, carrying 20 astronauts for a total of more than 40 days in space.

READING SELECTION

APOLLO MISSIONS

The Apollo missions were among the most important technological feats in U.S. history. They included 11 piloted missions, six of which ended on the Moon. Twelve Americans walked on the lunar surface, beginning with astronauts on *Apollo 11* on July 20, 1969.

The scientific investigations conducted during the Apollo missions changed our understanding of Earth, as well as our understanding of the origin of the Moon. Astronauts gathered samples of lunar soil and rock from the Moon and brought them back to Earth. These rocks and soils showed that the Moon is rich in resources needed to sustain human life—except for water, a vital ingredient for human existence.

The triumphs of Apollo were paired with tragedy, however. During a training accident in 1967, the first three Apollo astronauts were killed. *Apollo 13* suffered a near-disaster, but the program continued with four more successful missions.

Photographs taken by Apollo astronauts, which depicted Earth suspended in space, forever changed the way we view our home planet. The Apollo Project demonstrated that humans can live and work in space, and even on other worlds. It also demonstrated that technologies developed for the space program were capable of changing our lives forever.

SKYLAB PROJECT

Skylab was America's first space station. The lab measured about 36 meters (118 feet) long. In building it, scientists adapted elements from the Apollo program. The "lower" level housed the crew's quarters, the toilet, galley and dining area, as well as an experiment area with devices such as a rotating chair and a bicycle ergometer. The "upper" level, or forward compartment, was the storage area for food, water, experimental equipment, and spacesuits. The station had two large solar panels and an unattached airlock at one end. It also had a large, X-shaped structure that consisted of four solar arrays radiating from a central structure known as the Apollo Telescope Mount.

Skylab was launched on May 14, 1973, and the first crews arrived at the station on May 25. At the lab, astronauts studied and photographed Earth, and conducted experiments such as how extended stays in space affected the human body. In 1979, the space laboratory, emptied of equipment and passengers, burned up in the atmosphere over Australia.

THE APOLLO–SOYUZ TEST PROJECT

The Apollo-Soyuz Project demonstrated that the United States and the then-Soviet Union could work together in space. On July 15, 1975, Alexei Leonov—the first human to walk in space—and Valery Kubasov were launched into orbit aboard the *Soyuz 19*. The U.S. Apollo team—Donald Slayton, Vance Brand, and Thomas Stafford—lifted off seven hours later from the Kennedy Space Center in Florida. The Apollo crew docked with the Soviet craft. On July 17, the two teams shook hands and the crews spoke to the people on Earth via live television. On July 19, the teams went their separate ways. Two days later, the Soviets landed in Siberia. The American crew stayed in orbit until July 24, conducting additional experiments before splashing down in the Pacific Ocean.

SPACE SHUTTLE TO SPACE STATION

It wasn't until 1981 that NASA resumed its space flight efforts. At that time, it proposed building a space station and space shuttles to service it. The goal was to use the shuttles like airplanes, which could be flown again and again, rather than like rockets, which would be used once and thrown away. NASA built the first shuttle using unmanned solid rocket boosters that fell into the Atlantic Ocean. The boosters were towed back to shore, where they were loaded onto rail cars for a trip back to the factory for rebuilding. The orbiter was completely reusable, but its engines were fed from a huge external tank that had to be thrown away on every mission. In 1993, a completely reusable shuttle was developed.

Development of a successful space shuttle has not been all smooth sailing. One of the earlier shuttles, *Challenger*, was destroyed on its tenth mission to space in 1986. The crew of seven people was killed. A small piece of equipment, called the O-ring seal, had failed, causing parts of the shuttle to fall apart shortly after launch. This tragedy was followed by the 2003 destruction of another space shuttle, *Columbia*, when it exploded during re-entry to Earth. It had already completed 27 missions, and was scheduled for just a few more. While the first shuttle tragedy caused NASA to take a break from its shuttle program, the second one prompted retirement of the whole shuttle fleet. The old shuttle fleet is being replaced with new spaceships, such as *Orion*, to take humans to the Moon and Mars.

Despite the risks of space travel, space shuttles will continue to be used to lift heavy payloads into orbit, provide a lab for astronauts to carry out scientific research in space, and provide a platform for retrieving satellites. They will bring astronauts to and from the International Space Station—the largest space project in history to date. Through the space station, several nations including the United States are working to build a laboratory in space. Research aboard the station continues to answer basic questions about humans living and working beyond the planet.

SCIENTIFIC PROBES

NASA has launched a number of other significant unmanned scientific probes, including *Pioneer*, *Galileo*, *Mariner*, *Voyager*, *Cassini*, and *Messenger*. These spacecraft have explored the moons, planets, and other places in our solar system. NASA sent several spacecraft to investigate Mars. Earlier missions were conducted by the *Viking* spacecraft that carried stationary landers to Mars to record data. Later, *Pathfinder* carried the first roving robotic probe, named *Sojourner*, to Mars. A recent mission (Mars *Exploration* Rover) includes two roving probes named *Spirit* and *Opportunity*. The Hubble Space Telescope and other science spacecraft have enabled scientists to make a number of significant discoveries about the planets and "deep space."

Space exploration has changed the way we live our lives. In addition to the pioneering work that NASA has done on aerodynamics and space applications, satellites have helped change the way we communicate. And a new era has begun with private space missions now underway. A $20 million spacecraft called *Tier 1*, paid for by a wealthy sponsor, made its maiden flight in 2004. Private space missions may pave the way for space tourism, in which regular people could travel to outer space at a reasonable cost. The history of space exploration has been relatively brief—but it has had a large impact. ■

▶ **SPACE SHUTTLE *CHALLENGER* LIFTING OFF**
PHOTO: NASA Johnson Space Center

DISCUSSION QUESTIONS

1. How did the military in the U.S. and elsewhere help launch space-exploration programs?

2. How has our view of what is possible for humans in space changed over time?

READING SELECTION
EXTENDING YOUR KNOWLEDGE

PHOTO: NASA Marshall Space Flight Center

We have learned much about Earth's neighboring planets in the solar system. We've sent flyby spacecraft to photograph them, put orbiters around them for longer study, deposited landers on their surfaces, and flown probes through their atmospheres. But what have we learned about Earth as a planet using space technology?

Earth's vast oceans set it apart from the other planets. Liquid water surrounds its continents, which are covered by contrasting lush vegetation and desert landscapes. From space we can see frozen white polar caps—like those on Mars—that cover Earth's poles. Swirling clouds, flashes of lightning, and volcanic gases are evidence of an active atmosphere.

The presence of life on Earth is one of its unique features—and, maybe because we are alive ourselves, it is one of our first topics of conversation about what makes Earth special. Macro- and microscopic organisms are teeming on land and in water. One hint that life exists

on Earth can be detected from space—the electromagnetic noise caused by radios and TV broadcasts. But the living world is only one aspect of Earth.

EARTH SYSTEM ENTERPRISE

The key to a better understanding of Earth is to explore how its systems of atmosphere (air), geosphere (land), hydrosphere (water), and biosphere (life) interact with each other. And the best way to study all of these systems together is from space. Mission to Planet Earth—now called Earth System Enterprise (ESE)—was established in 1991 and is the foundation of NASA's Earth-observing program.

This program has three main components: a series of Earth Observing System (EOS) satellites, a system to collect data, and teams of scientists around the world who study the data. Through ESE, NASA hopes to explain how natural processes like continental drift, climate change, volcanic activity, and other phenomena affect life on Earth. ESE is also studying how life on Earth is affecting Earth's natural processes. This research may give us more accurate weather forecasts, help manage agriculture and forests, provide information about marine life to fishermen, and, eventually, help predict how the climate may change in the future.

Let's look at a few examples of how the ESE mission helps scientists and engineers observe Earth as a planet.

CLIMATE REGULATION BY CLOUDS

Clouds help to regulate Earth's climate. Cirrus clouds—high, wispy clouds—keep Earth warm

▶ THIS IMAGE IS A COMPOSITE FROM EIGHT DAYS OF REMOTE-SENSING BY TWO SPECTRORADIOMETERS, EACH MOUNTED ON AN EOS SATELLITE. ONE SATELLITE FLIES NORTH TO SOUTH ALONG THE EQUATOR IN THE MORNING, AND THE OTHER FLIES SOUTH TO NORTH IN THE AFTERNOON, PROVIDING A DIFFERENT PERSPECTIVE.

PHOTO: MODIS Land Group, NASA Goddard Space Flight Center

by trapping heat and light that leaves the Earth's surface. Stratocumulus clouds—soft gray clouds in globular patches or rolls—cool Earth's surface by reflecting incoming solar radiation back into space.

Scientists who work with Earth-observing satellites are studying how clouds affect Earth's climate. By using global cloud observations from EOS satellites, scientists can determine to what extent warming or cooling caused by clouds has an impact on the global climate. Sometimes they get unexpected results. For instance, following the terrorist attacks of September 11, 2001, most air traffic stopped. This meant that jets were not leaving contrails—white plumes of smoke—in the atmosphere. Contrails are thicker than clouds and block solar radiation, and it turns out that they help control the planet's temperature. Fewer contrails meant that temperatures on Earth rose and fell more than normal from day to day. Monitoring those temperatures gave us a rare view into how our activities directly affect the Earth's climate.

GLOBAL ICE AND SEA LEVEL CHANGES

Earth's glacial ice contains more than 77 percent of Earth's fresh water. Over the last century, many of the world's mountain glaciers and ice caps have been retreating (getting smaller). But it can be hard to tell how much ice is being lost, and how fast. Why? Because as the seasons come and go, cold areas gain and lose ice. One way scientists track ice loss around the world is by watching where the water goes: the oceans. When ice is lost, the oceans rise. So one of the jobs of the EOS scientists is to figure out whether Greenland and Antarctic ice sheets are growing or shrinking by studying changes in sea level. Scientists examine changes in sea level by looking at satellite, laser, and radar data.

GREENHOUSE EFFECT

Scientists use the EOS satellites to measure levels of greenhouse gases such as carbon

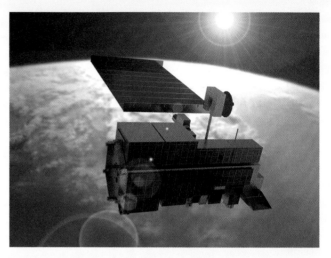

▶ *TERRA*, AN EARTH-OBSERVING SATELLITE

PHOTO: Barbara Summey, NASA, GSFC

dioxide, methane, and chlorofluorocarbons (CFCs) in our atmosphere. These gases trap heat within Earth's atmosphere, preventing it from escaping into space. Carbon dioxide is released into the atmosphere when solid waste, fossil fuels (oil, natural gas, and coal), and wood and wood products are burned. CFCs are found in aerosol sprays, in blowing agents for foams and packing materials, in solvents, and in refrigerants. CFCs are all manmade. If their levels increase in the atmosphere, it's entirely the result of human activity.

The levels of greenhouse gases have been increasing steadily. Knowing how fast they've risen is important. It lets policymakers around the world set goals for reducing greenhouse gas emissions.

OCEAN PROCESSES

Oceans transport heat around the globe, and we can see how they do this via satellite. An EOS satellite called *Terra* has collected detailed measurements of the ocean's surface temperatures every day all over the globe. Using a suite of sensors, *Terra* acts like a sophisticated thermometer in space. With the data, we can build temperature maps of

▶ THIS PHOTO OF SAN QUINTÍN GLACIER IN SOUTHERN
CHILE WAS TAKEN BY THE CREW OF MISSION STS-068 IN
OCTOBER, 1994.

PHOTO: Image courtesy of the Image Science & Analysis Laboratory, NASA
Johnson Space Flight Center (STS068-260-73, http://eol.jsc.nasa.gov)

▶ THIS PHOTO OF SAN QUINTÍN GLACIER WAS TAKEN ABOUT
SIX YEARS LATER, IN MAY 2000. LIKE MANY GLACIERS
WORLDWIDE, SAN QUINTÍN APPEARS TO BE RETREATING.

PHOTO: NASA/GSFC/MITI/ERSDAC/JAROS, and U.S./Japan ASTER Science Team

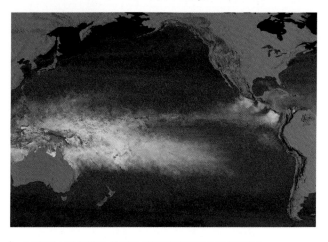

▶ A SATELLITE VIEW OF SEA SURFACE TEMPERATURES
IN THE PACIFIC OCEAN. COLD WATERS ARE BLACK
AND DARK GREEN. PURPLE, ORANGE, AND YELLOW
REPRESENT PROGRESSIVELY WARMER WATER.

PHOTO: NASA Goddard Space Flight Center Scientific Visualization Studio

the oceans' surface, assigning colors to each temperature range. And with those maps we can see where the warmest and coolest ocean surface waters are. Why is that important? Among other reasons, regions of warm and cool ocean water help drive the weather on Earth, as well as the climates of different areas on land. Scientists can use satellite imagery to study El Niño—an occurrence of unusually warm surface water in the Pacific Ocean.

EOS can also help scientists investigate the role Earth's oceans play in regulating the amount of greenhouse gases in the atmosphere. As these gases build up in the atmosphere, they are also absorbed into the oceans. This can have profound effects on marine life. Scientists are finding, for instance, that the infusion of carbon dioxide makes the ocean more acidic, which can lead to trouble in the food chain. The acidic environment weakens the shells of tiny marine creatures that are the base of ocean food webs. Any impacts to their populations could affect all the organisms that feed on them.

OZONE, VEGETATION, AND SNOW

For decades, satellite and ground-based measurement tools have tracked a growing hole in the ozone over the Antarctic. EOS analyzes the natural and human activities on Earth that

READING SELECTION

EXTENDING YOUR KNOWLEDGE

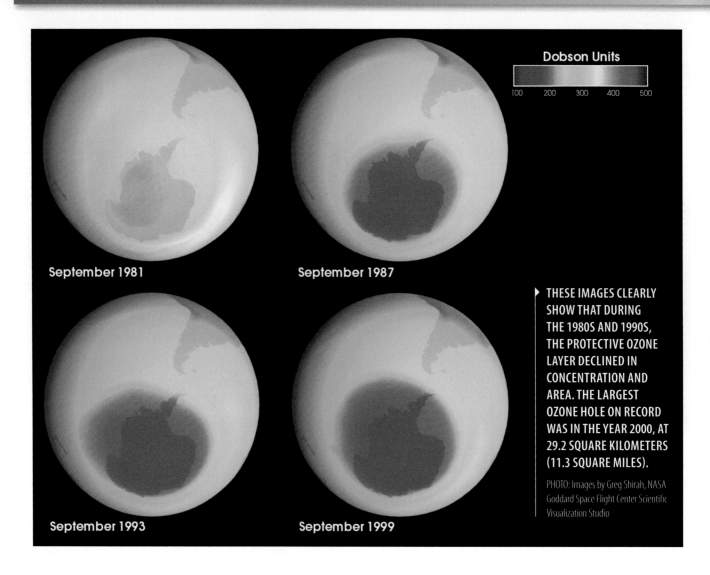

Dobson Units

100 200 300 400 500

September 1981

September 1987

September 1993

September 1999

▶ THESE IMAGES CLEARLY SHOW THAT DURING THE 1980S AND 1990S, THE PROTECTIVE OZONE LAYER DECLINED IN CONCENTRATION AND AREA. THE LARGEST OZONE HOLE ON RECORD WAS IN THE YEAR 2000, AT 29.2 SQUARE KILOMETERS (11.3 SQUARE MILES).

PHOTO: Images by Greg Shirah, NASA Goddard Space Flight Center Scientific Visualization Studio

cause the decrease in the ozone layer. The ozone is a layer of O_3 in the stratosphere, the second layer of the atmosphere above Earth's surface. Normally, our ozone layer protects us from some of the Sun's ultraviolet (UV) radiation. Ultraviolet radiation can harm organisms, including humans, by damaging cells' DNA, which can cause cancers and other genetic illnesses. The hole in the ozone layer allows UV radiation in freely. So in Australia, which is close to the ozone hole, school children are required to wear hats while outside. People are told to "slip, slop, slap" in order to avoid skin cancer: slip on a shirt, slop on some sunscreen, and slap on a hat.

Scientists at NASA also monitor processes that directly affect Earth's energy and water cycles. For example, satellites monitor the rate of deforestation (the process of cutting down trees) in the Brazilian rainforest. Satellite images can also show characteristics of vegetation: for instance, how much sunlight the leaves are absorbing.

Satellites monitor snow and ice cover at the poles to understand how global temperatures are changing. A decrease in the amount of ice at

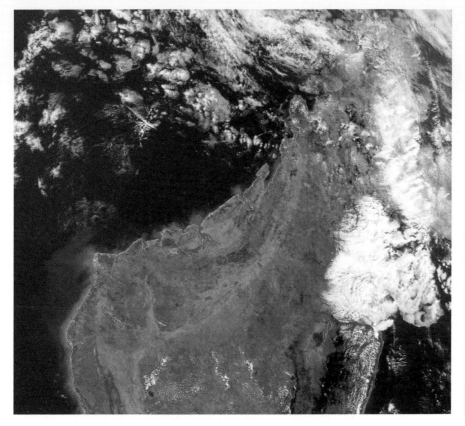

MADAGASCAR WAS ONCE COVERED IN LUSH GREEN VEGETATION. TODAY, AN ESTIMATED 80 PERCENT OF ITS FORESTS HAVE BEEN DESTROYED. THE REDDISH-BROWN EXPOSED TERRAIN CAN BE SEEN IN THIS TRUE-COLOR IMAGE OF NORTHERN MADAGASCAR, TAKEN IN MAY 2000.

PHOTO: Image by Brian Montgomery, Robert Simmon, and Reto Stöckli, based on data provided by the MODIS science team

the poles, for instance, indicates global warming, a matter of serious concern. Ice reflects light, which helps keep the planet cool. As sea ice melts, sunlight hits open ocean water, which is dark and absorbs the sunlight. This warms the oceans and speeds up the melting of ice in a vicious cycle. Monitoring how fast this process goes tells us about the urgency in dealing with global warming.

THE FUTURE OF EARTH AS A PLANET

So where do we stand? Earth is a system that harbors life because of the balances between water in its three states (liquid, gas, solid), temperatures, and oxygen levels. Our technology teaches us about and helps us care for both the living and non-living parts of this system that we rely on. The technology of the Earth System Enterprise mission promotes the study of Earth as an integrated system. ■

DISCUSSION QUESTIONS

1. Why do we need a space mission in order to study Earth?

2. What individual actions of ours end up having global environmental impacts? How do they become global?

PLANETARY FACTS

EARTH

EARTH: QUICK FACTS*

▶ **Diameter**
12,742 km

▶ **Average distance from the Sun**
149,598,262 km

▶ **Mass**
597×10^{22} kg

▶ **Surface gravity (Earth = 1)**
1[†]

▶ **Average temperature**
15°C

▶ **Length of sidereal day**
23.93 hours

▶ **Length of year**
365.26 days

▶ **Number of observed moons**
1

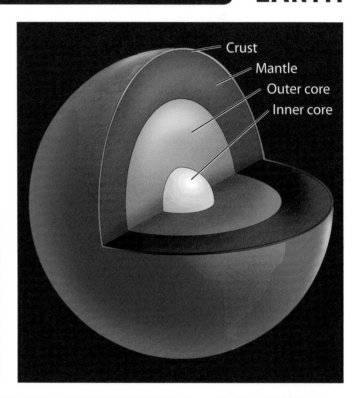

Crust
Mantle
Outer core
Inner core

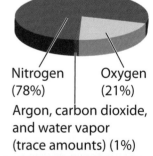

RELATIVE SIZE

Moon

DID YOU KNOW?

▶ The oldest rocks on Earth date back 4 billion years.

▶ Only Earth has the temperature range that permits liquid water to exist, and only Earth has developed an oxygen-rich atmosphere. These two factors enable Earth to support life.

ATMOSPHERE

Nitrogen (78%)
Oxygen (21%)
Argon, carbon dioxide, and water vapor (trace amounts) (1%)

Earth

* Source: Data from NASA as of 2011   † 9.81 m/s²

Climate's link to Life

Earth has a lot going for it—it has the best position of any planet in the solar system for maintaining life. If Earth were much closer to the Sun, all of its water would boil away, as it has on Venus. If our planet were farther from the Sun, Earth's water would freeze, as it has on Mars.

All the other planets in our solar system are either too hot or too cold to support life as we know it. But on Earth an incredible range of life forms exists, from algae to zebras. That doesn't mean, however, that Earth's climate is the same everywhere on the planet, or that it never changes.

CLIMATE DIFFERENCES IN PLACE AND TIME

Different climates support different plants and animals. Around the equator, thanks to nearly direct sunlight and warm seas, the climate is hot and wet, which is perfect for monkeys, snakes, and orchids. In the frigid, dry Arctic, however, life is less abundant and must adapt to low temperatures, weaker light, and long, dark winters. Can you imagine an orchid surviving in the Arctic?

Not only does Earth have a wide variety of climates, but over time, the planet's climate has changed. Throughout Earth's 4.5 billion-year history, there have been many ice ages, often lasting thousands of years, when snow and ice covered much of the planet. Eventually, the atmospheric and oceanic systems that produced those ice ages changed, putting an end to each ice age. But there were losers each time the climate changed.

During Earth's last ice age, woolly mammoths and mastodons thrived. Scientists have found

▶ THIS PHOTO OF PRINCE WILLIAM SOUND, ALASKA, SHOWS EARTH'S CONDITIONS IN A SUBARCTIC CLIMATE.

PHOTO: Commander John Bortniak, NOAA Corps, NOAA's America's Coastline Collection

READING SELECTION

EXTENDING YOUR KNOWLEDGE

▶ **DIFFERENT CLIMATES SUPPORT DIFFERENT PLANTS AND ANIMALS ON EARTH**

many of their bones and analyzed these fossils to determine their ages. The research makes one thing perfectly clear: these prehistoric giants died out at the end of the last ice age. They did not survive the change in climate. Their disappearance helped make way for newer species of plants and animals that were better suited to the warmer climate.

Thousands of years from now, Earth will likely enter another ice age. But the global concern now, as you probably know, is a period of warming.

GREENHOUSE GAS

Earth typically experiences a warm period between ice ages, but something more dramatic is currently happening on our planet. Evidence suggests that it is mainly due to the release of carbon dioxide into the atmosphere. Carbon dioxide is a "greenhouse gas;" it lets the Sun's heat in, and traps it in the atmosphere. Like a greenhouse, the atmosphere warms from the trapped sunlight.

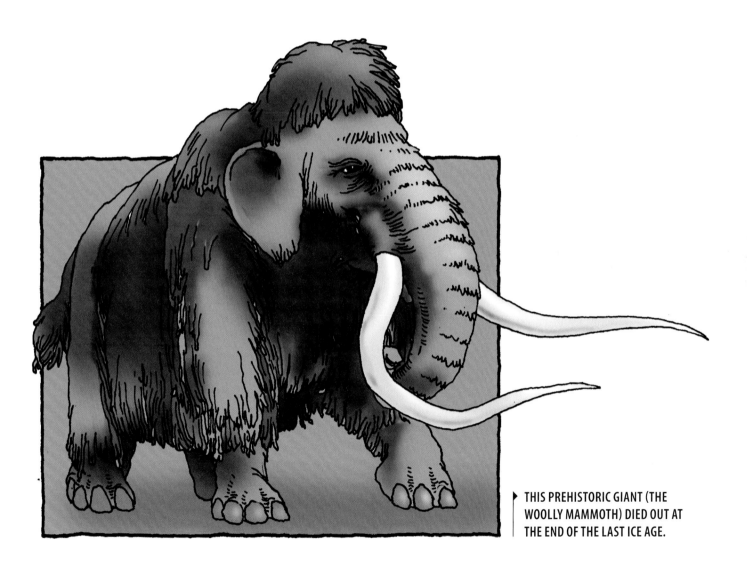

▶ THIS PREHISTORIC GIANT (THE WOOLLY MAMMOTH) DIED OUT AT THE END OF THE LAST ICE AGE.

READING SELECTION
EXTENDING YOUR KNOWLEDGE

▶ **SCIENTISTS ARE TRYING TO UNDERSTAND THE IMPACT OF GLOBAL WARMING ON OTHER SPECIES.**

PHOTO: Eric Regehr/U.S. Fish and Wildlife Service

Where is the release of carbon dioxide coming from? It is driven by our use of fossil fuels: carbon-based fuels like oil, coal, and natural gas. These rich carbon stores have been buried in the earth for millions of years. While burning them releases a great deal of energy, it also combines oxygen with the carbon, producing carbon dioxide (CO_2), a colorless gas. It's released into the air.

We are not the only force on the planet releasing carbon dioxide into the atmosphere. Every time plants burn in natural forest and grassland fires, their carbon combines with oxygen to form carbon dioxide. However, carbon dioxide released into the atmosphere on this scale is reabsorbed by living plants, which use the carbon in building their bodies. Unfortunately, the amount of carbon dioxide we send into the air by burning long-buried fuels cannot be absorbed by Earth's plants.

Atmospheric levels of carbon dioxide have risen, and, as they have, more carbon dioxide has also been forced into the oceans.

To make matters worse, we are rapidly chopping down forests in many parts of the world: using the space for housing and farming, and using the wood for fuel. Chopping down a forest tends to destabilize soil, allowing rich topsoil to be washed or blown away, and making it harder for new carbon-dioxide-absorbing forests to take root there.

GLOBAL WARMING

Temperatures on Earth are rising. Scientists predict that increased hot spells may cause even more drastic changes related to the heat. Numerous frog species already have been wiped out in Central America because of these changing conditions. Some scientists believe that the world's coral reefs may be gone by

2050 due to global warming. A prolonged warm period during the 1970s caused populations of Antarctic emperor penguins to drop by half. Lower temperatures had led to reduced availability of krill, the tiny shrimp on which the penguins feed.

Higher temperatures also will melt glaciers and ice sheets. Models indicate that the glaciers in Glacier Park, Montana, may be gone by 2030; around the world, snow-covered mountains are no longer snow-covered. In steppe and tundra biomes, permafrost—permanently frozen ground—has been melting, letting soils shift, sometimes leaving the few residents in those areas with a crazy landscape of buckled roads and tipped-over telephone poles.

One measure of global warming that scientists watch carefully is the size and density of the polar ice caps. These have been shrinking, leaving large new patches of open ocean water. Recently, scientists have observed dramatic demonstrations of how ice sheets melt. They've seen large "melt lakes" on top of ice sheets suddenly vanish as the ice beneath them cracks and allows millions of cubic feet of fresh, melted water to rush into the salty oceans below.

Melting ice leads to rising sea levels as the oceans receive the meltwater. Rising sea levels in Maryland's Blackwater National Wildlife Refuge are driving away many species of birds. Coastal areas around the planet—including places such as Florida in the United States, and countless low-lying countries like the Netherlands—could eventually become covered with water.

These changes are cause for real concern. A shift of just a few degrees can radically change our world's climate, the way we live, and perhaps our ability to survive. The most serious fears center on what are called "positive feedback loops": changes that encourage more of the same kind of change. Sea ice melt, for instance, leaves patches of open water that absorb more of the Sun's heat than ice does. This in turn heats Earth, leading to more sea ice melting. How fast and how far the changes will go is still a matter of intense debate, but there is consensus that governments around the world should try to reduce carbon dioxide emissions.

We are all interested in keeping Earth comfortable for human life. If we fail, of course, we'll have a difficult time ahead. But Earth is home to a great diversity of creatures, some of which thrive in extreme heat. It could be that a warmer Earth would be home to many creatures—just not to the ones that we depend on. ■

DISCUSSION QUESTIONS

1. What observations led scientists to conclude that Earth is warming?

2. Why does it matter if ocean levels rise and the oceans warm?

IMPACT CRATERS

INTRODUCTION

Impact! Suddenly, an enormous explosion occurs. The tremendous heat generated by the speed of the incoming asteroid causes the rock to vaporize upon impact. The impact creates a deep round hole.

Impact craters—bowl-shaped holes in the surface of a rocky planet, moon, or asteroid—are the most widespread landform in the solar system. Earth has not been spared from asteroid impacts, although the effects of water and wind erosion, volcanism, and other processes on Earth have erased much of the evidence. On our Moon, and on planets like Mercury where there is no atmosphere, craters remain intact and therefore may hold many secrets about the solar system's past.

What do craters look like and how do they form? In this lesson, you will examine photos of craters. You will model the formation of impact craters and design an investigation to test the factors that affect the size and shape of craters formed during an impact. You also will read about Mercury.

▶ **WHAT WOULD HAPPEN IF A LARGE ASTEROID COLLIDED WITH EARTH?**

PHOTO: Don Davis, NASA

OBJECTIVES FOR THIS LESSON

▸ Classify photographs of planets, moons, and asteroids on the basis of their surface features.

▸ Model the effects of impact cratering.

▸ Design an experiment to investigate impact craters.

▸ Summarize and organize information about Mercury, and compare Mercury with other planets.

▸ **MATERIALS FOR LESSON 5**

For you

   1   working copy of Student Sheet 2.3c: Solar System Chart

   1   pair of goggles

   1   pair of red and blue 3-D stereo glasses

For your group

   1   copy of Student Sheet 5.2a: Planning Sheet

   1   copy of Student Sheet 5.2b: Recording Our Crater Data and Conclusions

   1   plastic box filled with the following:

         Sand

         Flour

         Cocoa

   1   large sifter cup filled with extra cocoa

   1   large sifter cup filled with extra flour

   1   large resealable plastic bag filled with the following:

     2   hand lenses

     1   plastic spreader

     1   pack of 3 steel spheres

     1   metric measuring tape

     1   ring magnet

     1   flashlight

     2   D-cell batteries

     1   metric ruler, 30 cm (12 in)

     1   protractor (optional)

     Newspaper

     Paper towels

GETTING STARTED

1 Look carefully at the images shown in Figures 5.1–5.6. What observations can you make about them? How are the images alike? How are they different? Classify the photos. Record your classifications in your science notebook. Be prepared to justify your classifications. 🖗

2 Share your observations and classifications with the class.

3 Discuss these questions with your group or class, as instructed.

  A. What do you notice when you look at the close-up view of the craters?

  B. How old do you think the crater in Figure 5.2 is?

  C. Do you think Earth has a lot of craters like this? Why or why not?

  D. How is Figure 5.5 similar to or different from the other images?

4 Brainstorm with the class on what you already know about craters and what you want to learn.

5 Record a description of what you think causes craters and where you think they are most likely to occur. Share your description with your partner.

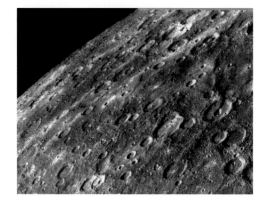

▶ **MERCURY**
FIGURE **5.1**

PHOTO: National Space Science Data Center, NASA/Image processing by United States Geological Survey

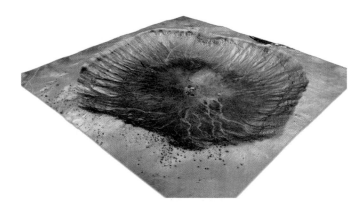

▶ **EARTH'S BARRINGER (METEOR) CRATER IN ARIZONA**
FIGURE **5.2**

PHOTO: NASA Goddard Space Flight Center Scientific Visualization Studio

▶ **THE "NEVER-SEEN-FROM-EARTH" FAR SIDE OF THE MOON**
FIGURE **5.3**

PHOTO: NASA Johnson Space Center

▶ JUPITER'S MOON, CALLISTO
FIGURE **5.4**

PHOTO: NASA Jet Propulsion Laboratory

▶ THE GIANT GASEOUS PLANETS—JUPITER, SATURN,
URANUS, AND NEPTUNE—ARE ALSO KNOWN AS
THE JOVIAN PLANETS.
FIGURE **5.5**

PHOTO: NASA/Lunar and Planetary Institute

▶ ASTEROIDS MATHILDE, GASPRA, AND IDA
FIGURE **5.6**

PHOTO: NASA

6 Look at Figure 5.7. How does it compare with the craters shown in Figures 5.1–5.4 and 5.6? How do you think this feature was formed? Discuss these questions with your group and then share your ideas with the class.

▶ SAN BENEDICTO ISLAND, MEXICO
FIGURE **5.7**

PHOTO: Lieutenant Debora Barr, NOAA Corps, NOAA's Small World Collection

INQUIRY **5.1**

MAKING GENERAL OBSERVATIONS ABOUT IMPACT CRATERS

PROCEDURE

1 Review the Safety Tips with your teacher.

SAFETY TIPS

Work in a well-ventilated area to minimize levels of dust in the air.

Wear indirectly vented goggles at all times during the inquiry.

Do not throw or project the metal spheres. Carefully release the metal spheres onto the powdered surface.

Do not stand on furniture to drop the metal spheres. Instead, place the box on the floor when testing drop heights greater than 60 cm.

2 Cover your workspace with newspaper. You may also want to protect your clothing. Obtain your group's plastic box, the large resealable bag of materials, and sifter cups of cocoa and flour.

3 Work with your group to investigate how craters form. Select one of the metal spheres. Drop it into the box. Use the magnet to carefully remove the metal sphere from the box without disturbing the surrounding powder. Observe the crater that it forms. Use your hand lens. Discuss what you see with your group.

4 Repeat Step 3 several more times. Use a new part of the box each time. Observe the crater that forms each time.

5 Try dropping the sphere from a different height or rolling the sphere along your ruler at different angles. Or try using a different size sphere. Observe your results each time. If the surface becomes covered in craters, carefully smooth out the powder with the edge of your plastic spreader (see Figure 5.8). Use the sifter cup to shake a layer of extra flour over the surface. Then sprinkle a thin layer of cocoa on the flour (the cocoa needs to be added on top of the white layer).

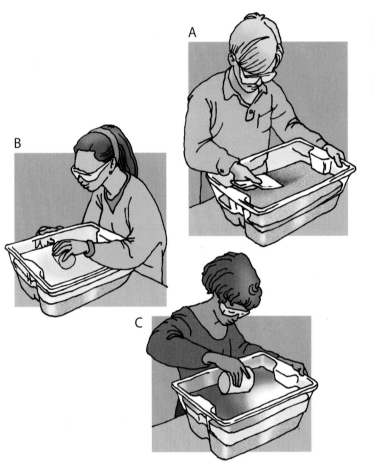

A

B

C

6 Share your general observations with the class.

7 Read "Craters in the Making" on pages 84–85 to find out more about crater formation. Discuss the parts of the crater with your class, referring to the illustration in the reading selection. What makes up the rays?

▶ WHEN YOU DO NOT HAVE ANY MORE ROOM TO CREATE CRATERS, (A) USE YOUR PLASTIC SPREADER TO SMOOTH OUT THE SURFACE OF YOUR FLOUR. (B) COVER THE SURFACE WITH A NEW LAYER OF FLOUR. (C) COVER THE FLOUR WITH A THIN LAYER OF COCOA.

FIGURE **5.8**

READING SELECTION

BUILDING YOUR UNDERSTANDING

CRATERS IN THE MAKING

A crater is a large, bowl-shaped hole found in solid, rocky surfaces. Terrestrial planets, moons, and asteroids all have craters. Some craters form when meteoroids, asteroids, or comets smash into the surface of a planet or moon. These craters—called impact craters—are usually circular. They range in size from tiny pits to huge basins hundreds of meters across. Impact craters are the most common geological features in the solar system.

THE IMPACT

The appearance of an impact crater depends on many things. For example, the size and speed of the object causing the crater affects its width and depth. With smaller or slower impacts, the surface material is simply thrown out, like sand that is thrown when a rock hits the surface of a beach. However, when the impacting object is large and traveling at higher speeds, it hits the surface with enormous force. The extreme temperatures and pressures from the collision cause the object to melt and mix with the surrounding rock.

PARTS OF A CRATER

After impact, a crater forms with a high rim and central peak (see the illustration at right). Landslides may create terraces around the rim. The floor of the crater is often below the level of the surrounding terrain. Ejected materials, such as dust, sand, and—if the temperature is high enough—liquid rock, fall back around the crater to form an area of debris that looks like spokes on a wheel. These spokes, called rays, radiate outward in all directions. Some rays can go for hundreds of kilometers beyond the point of impact. Many smaller, secondary craters also form around the main crater.

Craters on moons and planets can be seen most easily when long shadows are cast on their surfaces. Craters along the border between the Moon's dark and light side are most visible from Earth when the Moon is in its quarter phase.

CRATERS: LINKS TO THE PAST

Craters can tell us about a planet or moon's history. The more craters on a planet or moon's surface, the older that part of the object's surface is. During the early formation of our solar system, many meteoroids bombarded the planets. The craters they caused can still be seen on the Moon and Mercury. This is because geological processes, such as wind and water erosion, stopped millions of years ago on these bodies (see Figures 5.1 and 5.3). The craters remain much as they were at the time of their creation.

Gaseous planets have little or no evidence of impact craters, even though meteoroids strike gaseous planets as often as they strike rocky planets. Craters leave only a temporary record in the gaseous atmosphere.

CRATERS ON EARTH

Earth also was heavily cratered during its formation, and it still receives impacts today. Many craters on Earth have been eroded by wind and water and destroyed by earthquakes and volcanism. More than 120 impact craters have been identified on Earth. Some of Earth's craters are relatively young; for example, Barringer (Meteor) Crater in Arizona (see Figure 5.2) is only around 50,000 years old. Manicouagan Crater in Quebec, Canada, is much older; it was created about 214 million years ago. At 70 km in diameter, Manicouagan Crater is one of the largest impact craters on the surface of Earth. ■

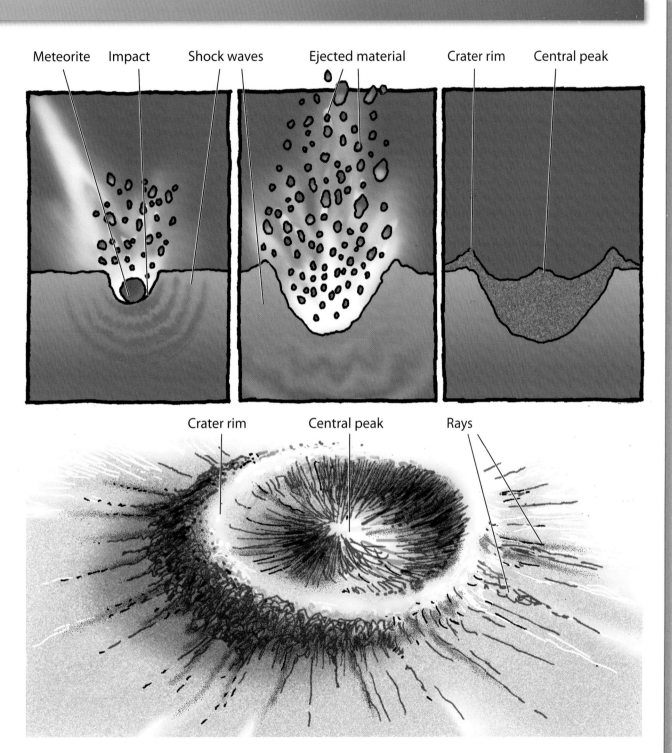

Meteorite Impact Shock waves Ejected material Crater rim Central peak

Crater rim Central peak Rays

▶ STAGES IN THE FORMATION OF IMPACT CRATERS INCLUDE THE IMPACT, THE BREAKUP AND MELTING OF THE IMPACTING
OBJECT, THE EJECTION OF BROKEN AND MELTED ROCK, AND THE FORMATION OF RAYS.

INQUIRY **5.2**

INVESTIGATING IMPACT CRATERS

PROCEDURE

1 Decide as a group what question about impact craters you will investigate. Try to choose a topic that might not be selected by another group. (If you have trouble, your teacher will review some questions your group can answer using your materials from Inquiry 5.1.)

2 Use Student Sheet 5.2a: Planning Sheet to design your group's crater investigation. Your teacher will approve your plan.

3 Decide how you will record your data.

4 Review the Safety Tips with your teacher.

5 Cover your work surface with newspaper.

6 Conduct your investigation. Remember the following points as you work:

- Change only the independent variable—the variable you are testing (for example, the height from which you drop your steel sphere).

- If you vary the drop height (which changes the velocity of impact), use even increments such as 30, 60, or 90 cm. Start with the smallest measurement.

- Use the ring magnet to remove the metal sphere. Avoid disturbing the crater.

- Use your hand lens to observe the details of your crater.

- Use tools to measure your dependent variables (for example, crater depth).

- If your surface becomes covered in craters, use the plastic spreader to smooth the surface. Use the sifter cups to create new layers of flour and cocoa.

- Record your data and conclusions in your science notebook or Student Sheet 5.2b: Recording Our Crater Data and Conclusions, as instructed. If you conduct multiple trials, calculate your average results.

SAFETY TIPS

Work in a well-ventilated area to minimize levels of dust in the air.

Wear indirectly vented goggles at all times during the investigation.

Do not throw or project the metal spheres. Carefully release the metal spheres onto the powdered surface.

Do not stand on furniture to drop the metal spheres. Instead, place the box on the floor when testing drop heights greater than 60 cm.

7 When every group has completed its investigation, your teacher will turn off the classroom lights. Use your flashlight to examine your craters (see Figure 5.9). Shine the light at the craters from all directions. From which direction are the crater features best seen? How does a flashlight change the appearance of the craters?

▸ USE YOUR FLASHLIGHT TO EXAMINE THE CRATERED SURFACE.
FIGURE **5.9**

▸ THIS APOLLO IMAGE SHOWS THE MOON'S KING CRATER IN 3-D.
FIGURE **5.10**

REFLECTING
ON WHAT
YOU'VE DONE

1 Work with your group to determine what conclusions you can make about impact craters on the basis of your evidence. Share your data and conclusions with the class.

2 Revisit the photographs in Figures 5.1–5.4 and 5.6. Answer the following questions in your science notebook, and then discuss them with the class:

A. How are the craters in the photographs like the craters in your plastic box?

B. Can a crater's appearance help you understand how the crater formed? Explain.

C. Where is the impacting object in each photograph?

D. Think back to when you used your flashlight. When is the best time for scientists to observe craters?

3 Get yourself a pair of red and blue glasses. The red lens goes over your left eye and the blue lens goes over your right eye. Use the glasses to observe the crater features shown in Figure 5.10. Give your eyes 30 seconds to adjust to the glasses. Can you observe any new features in the crater? If so, make a record of them in your science notebook.

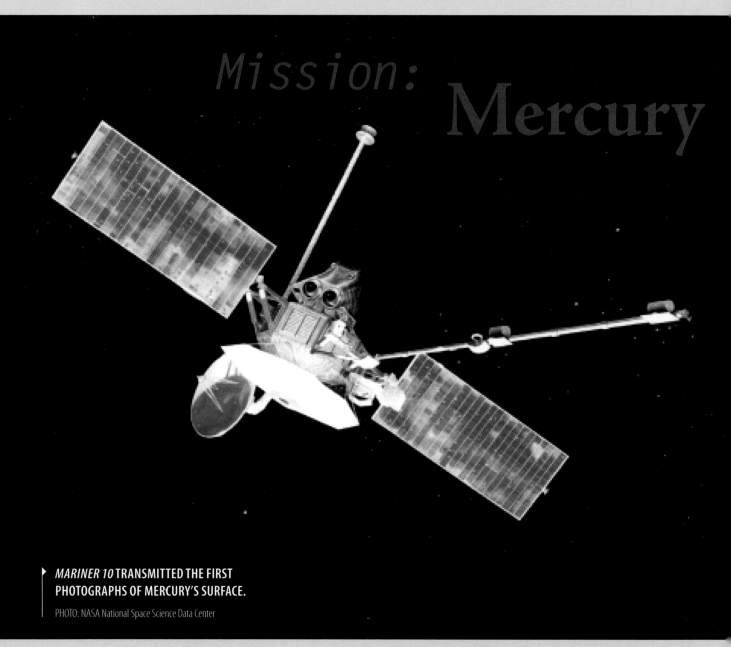

Mission: Mercury

▶ *MARINER 10* TRANSMITTED THE FIRST
PHOTOGRAPHS OF MERCURY'S SURFACE.

PHOTO: NASA National Space Science Data Center

Five thousand years ago, astronomers in Sumeria (located in present-day Iraq) identified Mercury and followed its wanderings across the sky. Because it is so close to the Sun, Mercury can only be seen along the horizon just before sunrise or just after sunset. When Mercury is directly overhead, the Sun's light obscures any view of the planet.

The fact that Mercury is difficult to view didn't stop scientists from trying to learn about it through the following millennia. At the turn of the 20th century, the astronomer Eugenious Antoniadi used a telescope to observe Mercury. He created maps of the planet that were used for nearly 50 years.

Scientists once thought that Mercury's day was the same length as its year because it didn't rotate at all, but merely zipped around the Sun in 88 days. But in 1965, Doppler radar observations were used to prove that Mercury rotates three times for every two times it orbits the Sun. Despite this new knowledge, scientists still had many questions about the innermost planet of our solar system. To learn more about Mercury, the Mariner 10 mission was launched from Kennedy Space Center on November 3, 1973.

THE *MARINER 10* FLIGHT

Mariner 10 was a small spacecraft. Its body was only 1.39 meters (4.56 feet) by 0.457 meters (1.499 feet)—less than the width of most classroom desks. However, solar panels, antennae, and sunshades attached to the body of *Mariner 10* added to its size. The spacecraft contained instruments to study the atmospheric, surface, and physical characteristics of Mercury, while the solar panels and rocket engine helped *Mariner 10* reach the planet.

Mariner 10 did not fly directly to Mercury. After leaving Earth's atmosphere, it flew by Venus and used that planet's gravitational pull to "slingshot" itself around Venus, bending its flight path toward Mercury. After almost five months in flight, *Mariner 10* made its first flyby of Mercury. The term "flyby" refers to a method astronomers use to observe a planet. Instead of having a spacecraft orbit or land on a planet, a flyby spacecraft does just that—it "flies by" a planet, taking pictures of it and gathering other scientific data. After *Mariner 10* completed its flyby of Mercury, it began orbiting the Sun. Using the last of its fuel, *Mariner 10* flew over Mercury two more times in September 1974 and March 1975 before becoming de-operational.

▶ A PHOTOMOSAIC OF HALF OF MERCURY'S SURFACE, CREATED USING IMAGES TAKEN BY THE *MARINER 10* SPACECRAFT.

PHOTO: NASA Jet Propulsion Laboratory

MARINER'S LOOK AT MERCURY

During each flyby, *Mariner 10* gathered images and data. Cameras on the spacecraft took more than 2,500 images of Mercury. These images mapped almost half of Mercury's surface. Why only half? Because Mercury rotates so slowly

READING SELECTION
EXTENDING YOUR KNOWLEDGE

that a single day on Mercury lasts longer than 58 Earth days! Because of the timing of the three flybys, the same side of the planet was always in the dark when *Mariner 10* was close enough to take photographs.

The photos taken by the *Mariner 10* revealed a heavily cratered surface much like that of the Moon. Craters with bright rays could be seen scattered among dark plains. These rays may be material ejected from the craters during impact. Scientists believe that some of Mercury's smooth plains—areas where craters appear to have been erased or covered over—were created by lava flow.

The cameras also took photos of a huge crater: Caloris Basin. This crater is 1300 kilometers (808 miles) across and may have been caused by an asteroid impact. Scientists believe this impact also caused the formation of hills on the other side of the planet. Other photographs taken by the *Mariner 10* showed crustal faults on the planet's surface. Evidence suggests that pieces of Mercury's solid crust have overlapped at these places.

The most surprising find of the Mariner 10 mission was that Mercury has a magnetic field. Some scientists think the magnetic field indicates that the planet has an iron core that

▶ SHOCK WAVES FROM THE IMPACT OF A LARGE METEORITE FORMED THESE CRATERS ON MERCURY.

PHOTO: NASA Jet Propulsion Laboratory

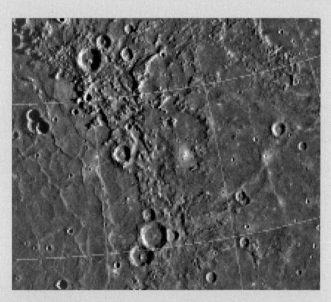

▶ THIS IMAGE, TAKEN BY *MARINER 10*, SHOWS ONE OF THE MANY FAULTS ON MERCURY'S SURFACE.

PHOTO: NASA Jet Propulsion Laboratory/Northwestern University is partially molten. Others think that an ancient magnetic field may be frozen in the crust.

MESSENGER VISITS MERCURY

Thirty years after *Mariner 10* flew past Mercury, NASA sent another probe, *Messenger*, to photograph Mercury and to explore its magnetic field, thin atmosphere, and surface features. *Messenger* was launched in 2004 but did not make its first flyby of Mercury until January, 2008. Data from the flyby indicated that the craters on Mercury are shallower than the

craters on the Moon by a factor of two, and that Mercury's iron core makes up 60 percent of its mass. That's twice as large as any other terrestrial planet. *Messenger* also found that Mercury has high cliffs, called lobate scarps, which formed when Mercury cooled and contracted.

Messenger made a second flyby of Mercury in October, 2008. On this flyby, *Messenger* photographed another 30 percent of the surface of Mercury that had never been seen before. On its third flyby, in September, 2009, *Messenger*

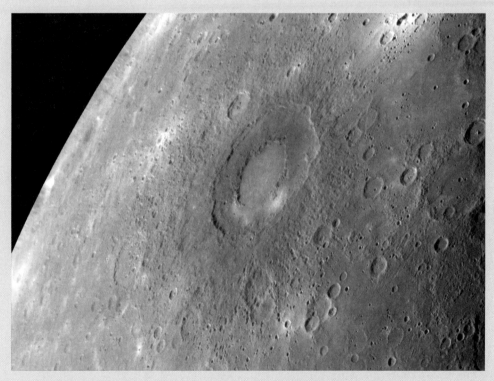

▶ THIS DOUBLE-RING CRATER ON MERCURY, FIRST PHOTOGRAPHED DURING THE THIRD FLYBY, WAS NAMED RACHMANINOFF, AFTER A FAMOUS RUSSIAN COMPOSER AND PIANIST, SERGEI RACHMANINOFF.

PHOTO: NASA/Johns Hopkins University Applied Physics Laboratory/Carnegie Institution of Washington

flew just 228 kilometers (142 miles) above the surface of the planet, giving it a gravity assist to enter orbit around Mercury. From this first-ever orbit of Mercury, scientists hope to learn more about its history and atmosphere.

Scientists also hope to land equipment on the planet, such as a camera, a seismometer, and tools for studying Mercury's soil. A spacecraft that lands on the planet is called a "lander." In such a mission, scientists hope to gather data directly from the planet's surface to answer numerous questions: What is the composition and structure of Mercury's crust? Has it really experienced volcanism? What is the nature of its polar caps? With future missions, perhaps we will have answers to these questions, and have a better understanding of this planet. ■

DISCUSSION QUESTIONS

1. What landscape features were found on Mercury, and how do they compare with features of the other planets you have studied during this unit?

2. What are the pros and cons of orbiters versus landers in terms of types of data they can obtain?

PLANETARY FACTS

MERCURY

MERCURY: QUICK FACTS*

- **Diameter**
  4879 km
- **Average distance from the Sun**
  57,909,227 km
- **Mass**
  33×10^{22} kg
- **Surface gravity (Earth = 1)**
  0.38
- **Average temperature**
  167°C
- **Length of sidereal day**
  58.65 Earth days
- **Length of year**
  87.97 Earth days
- **Number of observed moons**
  0

COMPOSITION

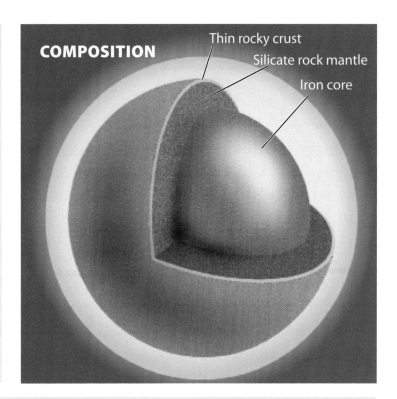

Thin rocky crust
Silicate rock mantle
Iron core

RELATIVE SIZE

DID YOU KNOW?

- Mercury is barely larger than Earth's moon. See the illustration "Relative Size" at left, comparing Earth to Mercury.
- Mercury was named for the messenger of the gods because it moves so quickly across the sky. It races along in its orbit at a rate of 46.4 kilometers per second—faster than any other planet!

Mercury
Earth

* Source: Data from NASA as of 2011

OTHER SURFACE FEATURES

INTRODUCTION

The most spectacular feature on Mars is the great canyon system called Valles Marineris. It extends for about 4000 kilometers. That's nearly a quarter of the way around the planet! The term canyon, however, is somewhat misleading. Why? On Earth, canyons generally are formed by running water. The canyons of Valles Marineris are basically cracks produced by tensions in the crust, although water is believed to have played a later role in shaping the canyons. After the cracks formed, deep springs seeped through the cliffs. This led to landslides, which were probably eroded by windstorms sweeping down the canyons. All of these processes combined to make Valles Marineris and its Ophir Chasm a spectacular planetary feature!

In this lesson, you will examine a set of photographs showing surface features on Earth. You will investigate whether these features exist on other planets as well. Various groups in your class will each model one of the planetary processes that create these features—wind erosion, water erosion, fractures caused by tectonics and other stresses, or volcanism. You then will match each group's model with photographs of different planets' surface features. The lesson ends as you read about three NASA missions to Mars.

▶ THIS IMAGE SHOWS PART OF THE OPHIR CHASM IN VALLES MARINERIS, A HUGE 4000-KILOMETER-LONG (2486-MILE-LONG) CANYON SYSTEM ON MARS. THE OPHIR CHASM IS SHAPED BY TECTONICS, WIND, SLUMPING, AND PERHAPS BY WATER AND VOLCANISM.

PHOTO: NASA Jet Propulsion Laboratory/U.S. Geological Survey

OBJECTIVES FOR THIS LESSON

Review photographs showing planetary surface features on Earth; then consider whether the processes that formed these features exist on other planets and moons.

Brainstorm what you know and want to learn about planetary processes on Earth and other planets.

Investigate wind erosion, water erosion, tectonics, and volcanism and their effects.

Analyze photographs of planetary surface features and determine how each was formed.

Summarize and organize information about Mars, and compare Mars with other planets.

▶ **MATERIALS FOR LESSON 6**

For you
- 1 working copy of Student Sheet 2.3c: Solar System Chart
- 1 copy of Student Sheet 6.1a: Planetary Process Observations (or your notebook)
- 1 copy of Student Sheet 6.1b: Matching Planetary Processes (or your notebook)
- 1 pair of goggles
- 1 pair of red and blue 3-D stereo glasses

For your group
- 1 large resealable plastic bag (from Lesson 5) filled with the following:
  - 2 hand lenses
  - 1 plastic spreader
  - 1 pack of 3 steel spheres
  - 1 metric measuring tape
  - 1 ring magnet
  - 1 flashlight
  - 2 D-cell batteries
  - 1 metric ruler, 30 cm (12 in)
  - Newspaper
  - Paper towels
- 1 set of four Planetary Process Photo Cards
- 1 set of process materials from the following:

  #### Wind erosion
  - 1 plastic box filled with sand, flour,

and cocoa (from Lesson 5)
- 1 sifter cup of all-purpose sand
- 1 sifter cup of flour
- 4 flexible straws

Water erosion
- 1 plastic box with drain hole and Velcro® filled halfway with dry sand
- 1 rubber stopper
- 1 cup with hole and Velcro®
- 1 bottle of clear tap water
- 1 bucket
- 1 large absorbent pad
- 1 small absorbent pad

Tectonics
- 1 plastic box filled with sand, flour, and cocoa (from Lesson 5)
- 1 sifter cup of sand
- 1 sifter cup of flour

Volcanism
- 1 plastic box filled with sand, flour, and cocoa (from Lesson 5)
- 1 piece of wide acrylic tubing
- 1 large plastic syringe
- 1 cup of flour, with lid
- 1 bottle of red water
- 1 plastic cup
- 1 plastic spoon

GETTING STARTED

1 Within your group, review the photos of surface features shown in Figures 6.1-6.5 and discuss the following questions:

- What observations can you make about each surface feature?

- How do you think the surface features shown in the photos were formed?

- Do you think the processes that created these features exist on other planets? Explain your answer.

▶ **RIPPLED SAND DUNES IN DEATH VALLEY, CALIFORNIA**
FIGURE **6.1**

PHOTO: Alan Van Valkenburg, National Park Service

DEBRIS FLOW IN LA CONCHITA, CALIFORNIA
FIGURE **6.2**

PHOTO: FEMA/John Shea

RADAR IMAGE OF GALERAS VOLCANO IN COLOMBIA
FIGURE **6.3**

PHOTO: NASA Jet Propulsion Laboratory

Getting Started continued

▶ **MARBLE CANYON AT THE NORTHEAST END OF THE GRAND CANYON IN ARIZONA**
FIGURE **6.4**

PHOTO: Mark Lellouch, National Park Service

▶ **SAN ANDREAS FAULT IN CALIFORNIA**
FIGURE **6.5**

PHOTO: R.E. Wallace, U.S. Geological Survey

2 Share your observations with the class.

3 Discuss what you already know about planetary surface features and what you want to learn about them. You may be asked to record your ideas in your science notebook. 📝

INVESTIGATING PLANETARY PROCESSES

PROCEDURE

1 Look at the materials list for this lesson. Which surface feature would your group like to model: wind erosion, water erosion, tectonics, or volcanism? Select one process, with your teacher's input.

2 Within your group, brainstorm ways that you might use the materials to model your selected process.

3 How will you record your observations? Discuss this with your teacher.

4 Review the Safety Tips with your teacher.

SAFETY TIPS

Work in a well-ventilated area to minimize the levels of dust in the air.

Wear indirectly vented goggles at all times during the investigation.

Do not throw or project the metal spheres.

Cover any work surface with newspaper to absorb excess water and to avoid slippery surfaces.

Inquiry 6.1 continued

5 Read the appropriate background selection ("Wind Erosion," pages 102-103; "Water Erosion," pages 104-105; "Tectonics," pages 106-107 or "Volcanism," pages 108-109) to learn more about the planetary process you have selected. You will need your red and blue glasses to view some of the images. (Put the red lens over your left eye.) Use the information in the reading selection to help you plan how to model that process.

6 Gather the materials you will need. Each set is labeled with the name of the process. Cover your workspace with newspaper before beginning. If you are testing wind erosion, volcanism, or tectonics, use the boxes of sand, flour, and cocoa from Lesson 5. If you are testing water erosion, use a stream table (a plastic box that has a drain hole and is filled halfway with sand).

7 Conduct your investigation using the background reading selection as your guide.

8 When you are finished with your investigation, hold a flashlight parallel to the table to observe the surface features in your box. Record your observations on Student Sheet 6.1a or in your science notebook, as instructed. Write what you did, what you observed, and why you think it happened. Record which photograph of Earth (Figures 6.1-6.5) most resembles your results. 🔎

REFLECTING
ON WHAT
YOU'VE DONE

1 Share your group's results with the class. With the classroom lights dimmed, again use your flashlight at "sunset" (parallel to the table) to show off the features of your new surface to other groups.

2 Once all groups have reported, get one set of Planetary Process Photo Cards. Review the four photo cards with your group. Discuss what you see. How do you think each surface feature was formed? Where do you think each photo was taken? Read the caption on the back of each photo.

3 With your teacher's guidance, go around the classroom to see other groups' results. Use your photo cards and Student Sheet 6.1b: Matching Planetary Processes (or your science notebook) to match each photo card to the results in each group's plastic box. How do you think these features were formed? Which feature on Earth (Figures 6.1–6.5) matches the feature shown on each photo card? Record your observations.

4 Discuss your findings with the class. Then clean up.

5 Did you know that the relative positions of surface features can help scientists decide the relative age of the surface feature? (For example, a crater on the surface of a lava flow shows that the crater is younger than the lava. However, if lava fills the crater, then the crater is older.) Look again at the photos in this lesson and Lesson 5. (You may even be able to use the computer program *Explore the Planets* to view additional images.) Can you tell whether each crater shown is younger or older than the land around it?

6 With your class, return to the Question C (What surface feature is shown? Where would you find this surface feature in the solar system?) and Question E (What processes created this landform? Does this landform exist on other planets or moons? Explain why or why not.) folder from Lesson 1 and their photo cards. Is there anything you would now change or add? Discuss your ideas with the class.

7 Read "Wet Like Earth?" on pages 110–112. Answer the following question in your science notebook.

A. Does water exist on Mars? Explain.

READING SELECTION

BUILDING YOUR UNDERSTANDING

WIND EROSION

PLANETARY WINDS

Wind is gas in motion. Wind can exist only on planets with atmospheres. Three of the terrestrial planets—Mars, Venus, and Earth—have atmospheres and therefore have winds. Mercury, the Moon, asteroids, and many of the moons of the gaseous planets do not have an atmosphere as we know it. This means they do not have winds.

The thinner (or less dense) the atmosphere, the faster the wind has to blow to make an impact on the planetary or lunar surface. It takes a powerful wind to move rock fragments on Mars, because its atmosphere is so thin. It takes very little wind to move rock fragments on Venus, which has a thick atmosphere. The density of Earth's atmosphere is somewhere between that of Mars and Venus. Streaks on a planet's surface caused by wind, like in the photo at right, are evidence that wind moves smaller particles around. Wind erosion happens when gas molecules bounce against the rocks and other surfaces. A dense atmosphere has a lot of gas particles. This means that a dense atmosphere can erode a surface faster than a thin atmosphere in the same amount of time.

MARTIAN WINDS

The Martian surface has been eroded by winds that swept away fine particles and left behind boulders. Boulder fields that were found at the *Viking 1* landing site on Mars resemble deserts on Earth.

You've seen sand dunes on Earth. But do sand dunes exist on Mars? Get out your red and blue glasses and examine the dramatic photo on the next page of dunes on Mars. This field of wavy dunes is found in Nili Patera, a volcanic depression in central Syrtis Major, the most noticeable dark feature on Mars—the "red planet."

WINDS ON VENUS

The upper atmosphere of Venus is very windy. The winds there reach speeds of up to 350 kilometers per hour. In the lower atmosphere, the wind speed decreases until it is nearly zero at the surface. The wind blows in the direction of the planet's rotation. Since Venus rotates very slowly, the Sun shines for a long time on the surface. As the Sun heats the surface of Venus, the warm surface also heats the air above it. The rising warm air may be responsible for Venus's winds. ■

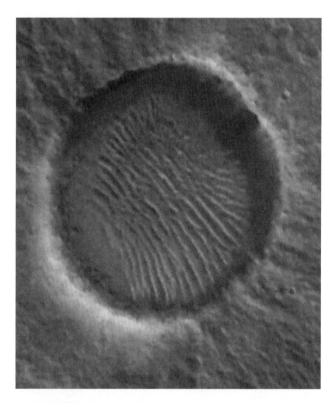

▶ **WIND STREAKS ARE EVIDENT IN A CRATER ON MARS**
PHOTO: NASA Jet Propulsion Laboratory/Malin Space Science Systems

3-D IMAGE OF SAND DUNES ON MARS. TWO DIFFERENT IMAGES FROM THE ORBITING MARS *GLOBAL SURVEYOR* SPACECRAFT WERE COMBINED TO MAKE THIS STEREO PICTURE. NOTICE THE RIPPLES ALONG THE SAND DUNES.

PHOTO: NASA Jet Propulsion Laboratory/ Malin Space Science Systems

MODELING WIND

You can use a straw and a box of sand, flour, and cocoa to model the effect of wind on a planet's surface (see the illustrations below). Do the following:

1. With your plastic spreader, smooth the layers of sand, flour, and cocoa in your plastic box. Don't worry if they mix.
2. Sprinkle a layer of sand on top of your mixture.
3. Sprinkle a thin layer of flour on top of the sand.
4. Use your flexible straw and the illustration as your guide. Blow very gently onto the surface of your plastic box. (Do not share straws.) What happens when wind blows over a fine dust? In which direction does the dust move? How do large and small particles change the effects of wind? Can you create dunes and wind streaks like those shown in the photos?
5. Use your flashlight to examine the wind streaks and sand dunes you have created. Remember that the best time to view planetary surface features is when the Sun is setting (when your light is horizontal and parallel with the table), not when it is directly overhead.

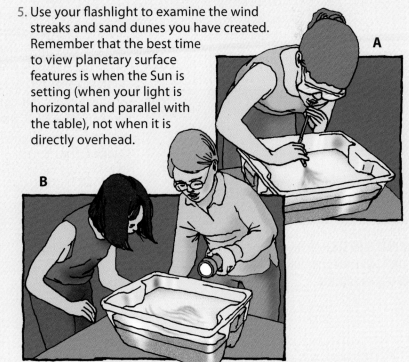

(A) USE A FLEXIBLE STRAW TO MODEL THE WIND'S EFFECTS ON A PLANET'S SURFACE. DON'T SHARE STRAWS. (B) USE THE FLASHLIGHT TO VIEW YOUR RESULTS. KEEP THE LIGHT PARALLEL WITH THE SURFACE OF YOUR WORKSPACE.

READING SELECTION

BUILDING YOUR UNDERSTANDING

WATER EROSION

A high mountain or plateau forms when forces within a planet lift up a relatively flat area. Flowing water cuts deep canyons into the highlands. Water typically flows from highlands to lowlands. The source of that water can include underwater springs, melting snow or ice caps, or rain. The flowing water erodes rock and creates canyons, valleys, and stream networks. Large boulders and small rock fragments may move with the water and help further erode the rock. Complex stream patterns often look like networks of nerves in the human body (see the illustration below). These patterns are common signs of water erosion on a planet or moon. A stream pattern depends on the slope, topography, soil type, and amount of water that flows across the surface.

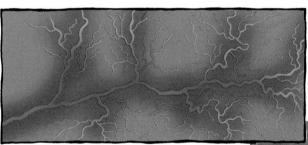

▶ STREAM PATTERNS SUCH AS THE ONE SHOWN HERE ARE COMMON SIGNS OF WATER EROSION.

WATER ON MARS

Running water formed the Grand Canyon on Earth, but Valles Marineris—a huge canyon on Mars—was formed by a combination of forces, mostly tectonics, although water is believed to have played a later role in shaping the canyons. After the cracks formed from tension in Mars's crust, deep springs seeped through the cliffs. This led to landslides. Valles Marineris is about as wide as the United States—some 4000 kilometers. The entire Grand Canyon could fit inside a small section of Valles Marineris.

How does water erosion affect existing craters on a planet's surface? Get out your red and blue glasses and examine this dramatic view of water channels on Mars. The channels, which were carved by water, are probably 200–300 meters deep. Water from one set of channels broke through a 12-kilometer-wide impact crater (center left in the photo below), and formed a large lake. The smooth floor may actually be lake sediment. ■

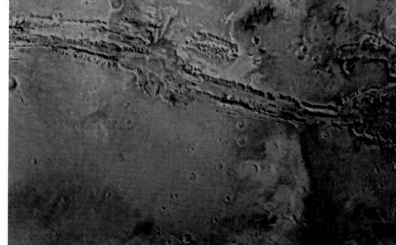

▶ VALLES MARINERIS

PHOTO: NASA Jet Propulsion Laboratory/U.S. Geological Survey

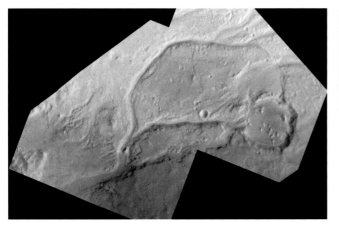

THIS *VIKING* 3-D VIEW SHOWS VEDRA VALLES, A GROUP OF NOW-DRY CHANNELS THAT ONCE FLOWED EASTWARD INTO THE GIANT CHRYSE BASIN ON MARS. THESE CHANNELS WERE CARVED BY RUNNING WATER WHEN MARS HAD LIQUID WATER ON ITS SURFACE.

PHOTO: © Lunar and Planetary Institute, 2000

MODELING WATER EROSION

You can use a stream table to model water erosion on planetary surfaces (see the illustration). Investigate water erosion by doing the following:

1. Place the large absorbent pad on your table with the plastic side face down over your newspaper. Place the small absorbent pad on the floor.
2. Position your "stream table" on the large pad so the drain hole hangs over the edge of your workspace.
3. Using the plastic spreader, push the sand away from the drain hole. Make the sand have a slope.
4. Remove the rubber stopper from the drain hole.
5. Hold the bucket directly under the drain hole, over the pad on the floor.
6. Attach the Velcro® on the cup to the Velcro on the stream table. Rock the cup back and forth until the cup is secured to the box.
7. Pour the water slowly into the cup. Try to keep the water at the top of the cup at all times. Do not touch the sand once you start to pour. Observe your results.

YOU CAN USE A STREAM TABLE TO MODEL WATER EROSION ON A PLANET'S SURFACE.

READING SELECTION

BUILDING YOUR UNDERSTANDING

TECTONICS

Tectonics is the study of how the outer layer of a planet can shift and break. Faulting occurs when parts of that outer layer move past one another. Folding (compression) occurs when parts of the outer layer collide and bend. Thinning occurs when the outer layer stretches. Jupiter's moon Ganymede (see the photo below, left) and Saturn's moon Enceladus both exhibit examples of faulting. Tremendous stress was created when large parts of the outer layer moved past each another. This buildup of stress eventually caused the rock to break along fault lines.

When the rocks on the outer layer of a planet collide, the rock folds. The compression stresses caused by those collisions can create wrinkle ridges in the surface of a planet or moon. Rock also can be stretched, which produces alternating large valleys and high standing blocks. These alternating blocks and valleys are evident on Mars and on Jupiter's moon, Ganymede.

Earthquakes occur along faults on Earth. But what do faults look like on other planets? Get out your red and blue glasses and examine the dramatic view of faults on Mars (below, right). Acheron Fossae is a set of valleys on Mars formed when the crust stretched and fractured. When two parallel faults form, the block of crust between them may drop down, with a ridge forming between them.

If an asteroid or comet hits a planet or moon, it can fracture its surface. Strong seismic waves result from the energy of the impact and move through the planet's or moon's surface. Fracture lines might radiate outward from the crater. Extreme temperature changes from day to night also can cause fracturing. (Think of how an ice cube fractures when you place it in a cup of hot water.) Fracturing due to extreme temperature change is especially common on planets or moons without an atmosphere. An atmosphere acts as a blanket that holds in heat at night and protects the surface from extreme high temperatures during the day. ■

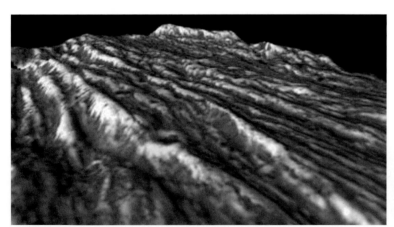

▶ **THE PATTERNS OF RIDGES AND GROOVES INDICATE THAT PULLING APART AND HORIZONTAL SLIDING HAVE BOTH SHAPED THE ICY LANDSCAPE OF JUPITER'S MOON, GANYMEDE.**

PHOTO: NASA Jet Propulsion Laboratory/Brown University

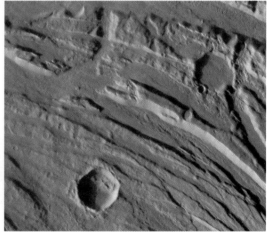

▶ **ACHERON FOSSAE IS A SET OF VALLEYS ON MARS.**

PHOTO: © Lunar and Planetary Institute, 2000

MODELING TECTONICS

You can model planetary tectonics by doing the following:

1. Use your large steel sphere and your plastic box of sand, flour, and cocoa from Lesson 5 to investigate the effects of impact cratering on a surface. Do fractures form around the crater?

2. Extreme changes in temperature, planetary shrinking, and other internal forces in a planet may cause it to twist or pull. Examine the effects of pulling and twisting forces on the surface of your box (see the illustration at right). Can you see cracks and wrinkle ridges forming in the surface as you twist and pull the sides of the box?

3. Now use your plastic spreader to push (compress) the layers of sand and flour (see the illustration below). Were you able to shift the layers? Do you see evidence of faulting?

▶ YOU CAN USE SAND, FLOUR, AND COCOA IN YOUR MODEL TO TEST THE EFFECTS OF PUSHING, PULLING, AND TWISTING TECTONIC FORCES ON A PLANET'S SURFACE.

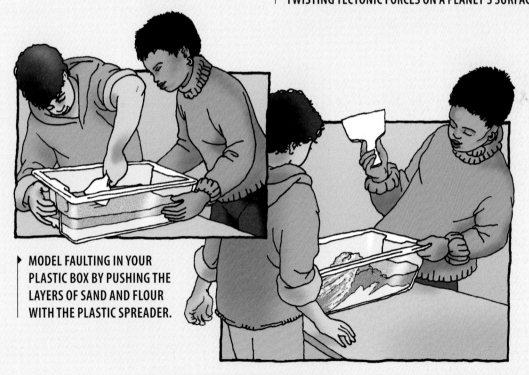

▶ MODEL FAULTING IN YOUR PLASTIC BOX BY PUSHING THE LAYERS OF SAND AND FLOUR WITH THE PLASTIC SPREADER.

READING SELECTION

VOLCANISM

Magma is made up of melted rock, crystals, and dissolved gases. It is found deep in a planet's interior. When magma erupts onto the planet's surface, it forms lava. This eruption creates distinctive landforms, such as lava plains and volcanoes.

VOLCANOES ON PLANETS AND MOONS

Dark, flat lava plains cover about 17 percent of the Moon's total surface. These plains are called "maria," from the Latin word for "seas." You can see the maria on the surface of a full moon on a clear night. The Moon's maria are made up of volcanic rock similar to the rock on Earth's ocean floor. The lava that formed the maria flowed long distances. Many maria were formed as lava flooded low-lying areas, such as the bottom of an impact crater. Older lava flows on the Moon have been covered by younger flows or have been dented with impact craters.

Jupiter's moon Io has numerous low, flat volcanoes called shield volcanoes. Most shield volcanoes have dark peaks with long lava flows streaming from them. The lava ran gently down the sides of the volcano. On Venus, the shield volcano Sif Mons resembles many of the shield volcanoes on Earth. The volcano Olympus Mons on Mars is the largest shield volcano in the solar system. It measures 600 kilometers across.

The volcanoes on Venus have unusual, circular, flat-topped domes called pancake domes. Wearing the red and blue glasses, examine the view of pancake domes on Venus in the photo below. The largest dome in this image is 65 kilometers across and roughly 1 kilometer high. This group of pancake domes is called Carmenta Farra. A small crater near the center of each dome may be the source for that dome's lava flow.

Why are shield volcanoes on other planets so large? Scientists believe that the crust of other planets, unlike Earth's crust, is not made of moving plates. Without plate tectonics, the volcanic vents on other planets may remain undisturbed for a long time. As a result, other planets have huge shield volcanoes many times larger than those on Earth. ∎

▶ 3-D VIEW OF VOLCANIC PANCAKE DOMES ON VENUS

PHOTO: © Lunar and Planetary Institute, 2000

MODELING VOLCANISM

You can model volcanism on other planets by doing the following:

1. Lava is often viscous (which means that it flows slowly). Mix red water and a small amount of flour in a cup to create a viscous model "lava." Don't make it too thick, or it will not flow through your tube.

2. Bury one end of your 90-cm tubing beneath the sand. Turn the end so it faces up.

3. Attach the unburied end of the tubing to the syringe (see the illustration below).

4. Pour lava into the syringe. Cap the syringe.

5. Very slowly compress your syringe to model lava eruption on your planetary surface. What do you observe? What happens when lava nears your craters?

▶ YOU CAN USE A SYRINGE AND LONG TUBING BENEATH THE SAND TO MODEL THE ERUPTION OF LAVA ONTO A PLANET'S SURFACE. TO CREATE YOUR LAVA, THICKEN YOUR RED WATER WITH FLOUR. WHAT WOULD A LAVA FLOW DO TO THE CRATERS ON THE PLANET'S SURFACE?

Wet Like Earth?

Scientists long suspected that Mars has plenty of water in the form of ice. Its large polar ice caps have been visible to people looking through telescopes here on Earth for many years—although it was not always clear that they were looking at frozen water rather than carbon dioxide and other frozen compounds. Scientists also have observed channels and valleys on Mars that indicate that water might have flowed across its surface long ago.

But there were plenty of arguments against the idea that Mars had ever had *liquid* water.

How could it have? Mars is colder than Antarctica and far drier than any place on Earth.

However, in June 2000, scientists at NASA announced an amazing discovery. New images transmitted from the Mars *Global Surveyor* spacecraft showed what looked like gullies on the surface of that planet. Gullies are ravine-like features that have been carved out of the crust's surface by flash floods. NASA scientists believed this to be evidence that Mars had flowing, liquid water in its recent past.

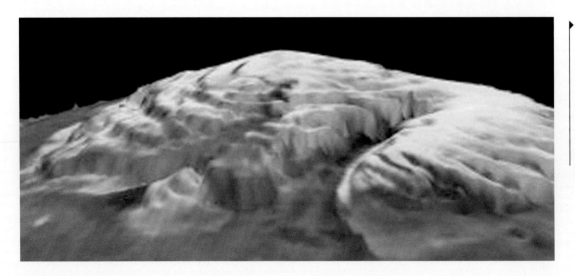

THESE POLAR ICE CAPS PROVE THAT MARS HAS PLENTY OF WATER IN THE FORM OF ICE.

PHOTO: NASA Jet Propulsion Laboratory/ Goddard Space Flight Center

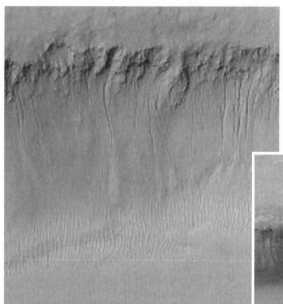

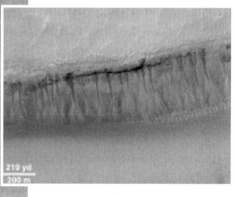

▶ SMALL GULLIES ON THE WALLS OF THIS VALLEY SYSTEM WERE CREATED WHEN A LIQUID—PROBABLY WATER—TRICKLED THROUGH THE WALLS UNTIL IT REACHED THE CLIFF, WHERE IT RAN DOWNHILL TO FORM THE CHANNELS AND FAN-SHAPED APRONS AT THE BOTTOM OF THE SLOPE.

PHOTO: NASA Jet Propulsion Laboratory/Malin Space Science Systems

▶ EVIDENCE OF LIQUID WATER IN A CRATER ON MARS, TAKEN BY THE MARS *GLOBAL SURVEYOR* ORBITER

PHOTO: NASA Jet Propulsion Laboratory/Malin Space Science Systems

STRONG EVIDENCE

The gullies in the NASA images are found on the sides of large craters or valley walls. Large pools of water, either from the surface of the planet or from beneath the surface, appear to have flowed from them. Areas of accumulated rocks and other debris at the lower ends of the gullies are evidence that if water was present, it flowed with great force.

The gullies have not been disturbed by wind erosion, asteroid impact, or volcanic activity. Scientists conclude that this means the gullies are extremely young, and that the water was liquid recently.

The idea that Mars had liquid water was supported by the work of the *Opportunity* rover in 2004. The robot explorer landed near the inside of a crater and began investigating the Martian soil. It found places in the rocks where crystals of salt minerals had formed salts which would likely have been deposited by the drying

of a watery environment. The chief investigator on the project, Dr. Steven Squyres, announced, "Liquid water once flowed through these rocks. It changed their texture, and it changed their chemistry."

And then—there was confirmation! In 2008, the *Phoenix* lander scooped up a soil sample, and along with it, chunks of an icy material. The ice evaporated, and the *Phoenix* lander confirmed that it was made of water, not carbon dioxide or another compound.

The news continues to come in about Martian water: The Mars *Reconnaissance* orbiter, launched in 2005, discovered in 2009 that Mars has large sheets of underground ice, nearly 99% water. These ice sheets are buried about halfway between the north pole and the equator. It's possible that future explorers on Mars could dig up this ice, melt it, and use it to sustain a human colony.

READING SELECTION

EXTENDING YOUR KNOWLEDGE

COLONIZING MARS

Why is the idea of liquid water on Mars so fascinating to astronomers? Well, liquid water in the past helps to support the idea that Mars once sustained life. But liquid Martian water in the present would allow people to live on Mars. Even though the Martian atmosphere is too thin to breathe, humans can make oxygen from water. They could also generate hydrogen for rocket fuel from water. And water means that humans could grow their own food.

There are certainly dangers. Mars's lack of a magnetic field means that the solar radiation reaching the surface may be too intense for humans. Also, Martian gravity is only 0.38 that of Earth's. Living in such low gravity for extended periods may cause health problems for humans.

Even so, there's considerable interest in sending people to Mars to set up a permanent camp there. One person who promotes the idea is Buzz Aldrin, the second man to walk on the Moon. His conversations about setting up a small colony on Mars have led others to adopt the slogan, "Forget the Moon, Let's Go to Mars!" Aldrin suggests we send 50 people to Mars by 2035, and have them conduct long-term research on the planet. It's a tantalizing idea; a round trip, Earth to Mars and back, would take about two years, not long for interplanetary journeys.

When people do reach Mars, they will begin to try to answer an important question—what happened to Mars? We need to know what caused the dramatic change in the Martian climate, which used to be so like our own. If we can find out, we might be better able to protect Earth.

THE MARTIAN RIDDLE

Whether life ever existed on Mars continues to intrigue scientists. A long time ago, conditions on Mars may have been similar to those on Earth today. The channels tell us that mighty Martian rivers may have once flowed into oceans. The atmosphere then may have been more dense and full of oxygen. Temperatures may have been much warmer. If so, life may have existed there and may exist there still.

Research on the Martian salt deposits may yield more clues about whether there has been life on Mars. Salt is an excellent preservative for materials from living things. Some 250-million-year-old salt deposits on Earth (from way below the ground in New Mexico) were found to contain dormant bacteria. The bacteria were sheltered in little pockets of saltwater within the crystals. Amazingly, scientists were actually able to revive and grow them. If bacteria have survived that long on Earth, might there be some life within the salt crystals on Mars? ∎

DISCUSSION QUESTIONS

1. What evidence suggests that Mars used to have liquid water on it?

2. Why are astronomers interested in whether or not liquid water existed on Mars? Why are they interested in whether water exists on Mars now?

Mission: Mars

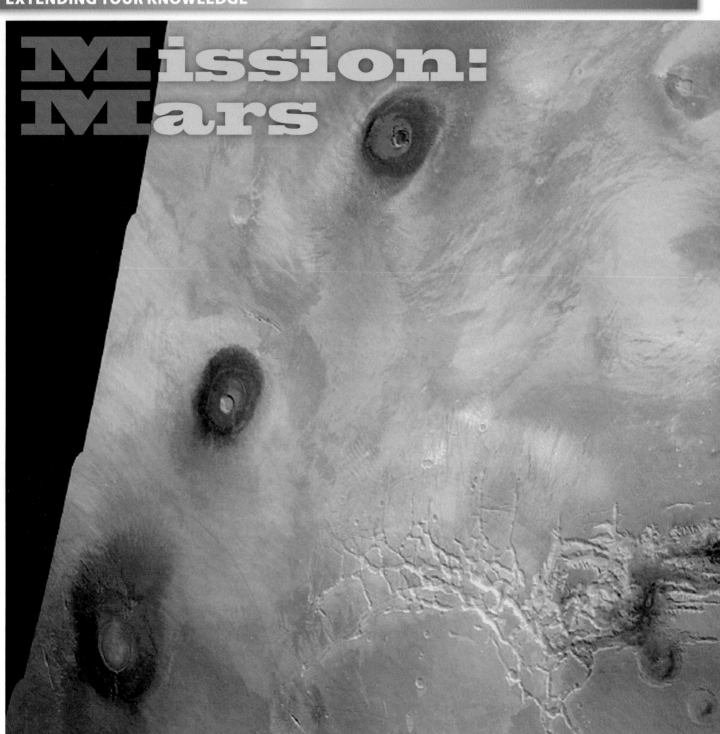

▶ A MOSAIC OF IMAGES TAKEN BY THE *VIKING 1* ORBITER SHOWS THE EASTERN THARSIS REGION ON MARS. NOTICE THE THREE VOLCANOES ON THE LEFT AND THE CANYON IN THE LOWER RIGHT-HAND CORNER.

PHOTO: NASA Jet Propulsion Laboratory

READING SELECTION

Mars's reddish color can be seen from Earth. So can its white polar caps. Some astronomers in the late 1800s saw dark lines running across the surface of the planet. Percival Lowell and other U.S. astronomers believed these lines were canals, built by Martians to transport water!

Not until 1965 would people know for sure what those lines were. In that year, *Mariner 4* flew by Mars and found a moonlike landscape, but no signs of life. Four years later, two other Mariner flyby missions confirmed these findings. Still, scientists believed that more time was needed to understand the red planet. That meant sending a spacecraft into orbit around Mars.

Mariner 9 was the first spacecraft to orbit another planet, although it had company within a few weeks of arrival: two Soviet orbiters, *Mars 3* and *Mars 4. Mariner* arrived at Mars in November 1971, and orbited the planet for nearly a year. During that time, it witnessed a month-long dust storm. It also found canyons, floodplains, and other signs of ancient water. Before the century was over, other American spacecraft—including *Viking 1* and *2,* the Mars *Pathfinder,* and the Mars *Global Surveyor*—would visit Mars.

VIKING 1 AND 2

Both *Viking 1* and *2* arrived at Mars in 1976. Each spacecraft was made up of an orbiter and a lander. The orbiter was designed to find a landing site for the lander, and to relay information from the lander to scientists on Earth. Although the two *Viking* spacecraft were designed to study Mars for several months, they actually provided scientists with data for several years.

Images from the orbiters showed volcanoes, lava plains, canyons, and craters on the surface of Mars. They also showed dry valleys and channels, which generated much excitement on Earth: perhaps Mars had once been a wet planet like ours. As expected, much of the landscape looked as if it had been carved by running water. The images also showed that Mars is divided into two main regions: northern low plains and southern cratered highlands.

With the orbiters flying overhead, the *Viking* landers descended through the thin Martian atmosphere. Their instruments revealed that carbon dioxide is the major gas in the Martian air. The two landers dropped down safely on opposite sides of the planet in iron-rich soil; it is iron that gives Mars its reddish color.

The landers' cameras searched the landscape for large forms of life—skeletons, perhaps, or even living creatures—but they didn't find any. Instruments on the landers conducted experiments to determine if the Martian soil at those sites contained microscopic life, kept alive, maybe, by scant water stores. But they found none.

The *Viking* landers did find a crusty surface that resembles Earth's crust. Tests also showed that Mars is extremely stable. There were no tremors, quakes, or volcanic eruptions.

READING SELECTION
EXTENDING YOUR KNOWLEDGE

MARS *PATHFINDER*

On July 4, 1997, the Mars *Pathfinder,* hidden inside a cocoon of huge air bags, bounced to a stop on the Martian surface. It had landed in an ancient plain where scientists believe a catastrophic flood left behind loads of rocks. Inside the *Pathfinder* spacecraft was a rover—a freely moving robotic explorer—named *Sojourner.* It would become the first rover to operate on another planet. *Sojourner's* task was to analyze Mars's rocks and soils.

Data also showed that Mars is drier and dustier than any desert on Earth. The dust often spins in gusts, called "dust devils." Clouds that cover parts of Mars consist of water ice condensed on the reddish dust.

Pathfinder's lander operated nearly three times longer than its expected lifetime of 30 days. The lander supplied photos of its position on Mars, and monitored weather with a variety of sensors. Its rover, *Sojourner,* had been expected to operate for a week, but instead went on exploring and relaying data for nearly three months. Together, they sent back 2.3 billion bits of information, including more than 16,500 images.

GLOBAL SURVEYOR

The Mars *Global Surveyor,* launched in November 1996, was equipped to fly at a low altitude in a nearly pole-to-pole orbit. This speedy spacecraft orbited Mars 12 times a day until 2006 when it went defunct.

Surveyor returned images of clouds hanging over gigantic volcanoes, dust storms that blow around the entire planet, and polar caps that enlarge in the winter and shrink in the summer.

▶ *SOJOURNER* IS TAKING A MEASUREMENT. NOTICE THE TWO-TONED SURFACE OF THE LARGE ROCK. WINDBLOWN DUST MAY HAVE COLLECTED ON THE SURFACE (THE ROCK IS LEANING INTO THE WIND). OR THIS ROCK MAY HAVE BROKEN OFF FROM A LARGER BOULDER AS IT WAS DEPOSITED IN THE ANCIENT FLOOD THAT SCOURED THIS AREA.

PHOTO: NASA Jet Propulsion Laboratory

Surveyor also confirmed that these polar caps consist of layers of dust and frozen carbon dioxide. Astronomers believe that such layers hold secrets to seasonal changes on Mars. They also may be one of the best places to search for evidence of past life on the planet.

Surveyor made another discovery that increases the chances of finding traces of past life. Mars, it turns out, once had a magnetic field, much as Earth does today. This is significant because magnetic fields shield planets from radiation that can harm the machinery of living cells. A magnetic field makes life possible where it otherwise would not be.

MORE MISSIONS

Astronomers continue to explore Mars and search for signs of life. In 2003 the Mars *Express* orbiter arrived at Mars. Mars *Express* was built and launched by the European Space Agency. The orbiter has photographed a system of outflow channels that suggest that lava deposition and floods have occurred on Mars. Mars *Express* has also observed aurorae on Mars: bright colors in the skies near the south pole, caused by solar particles striking a magnetic field generated by rocks in a certain area. This is very unusual, and at first it puzzled astronomers, because they knew that Mars currently has no planetary magnetic field. They suspect these small magnetic fields are actually remnants of the old, planetary magnetic field.

▶ THIS IMAGE SHOWS *GLOBAL SURVEYOR* ABOVE MARS.

PHOTO: NASA Jet Propulsion Laboratory

▶ TWELVE ORBITS A DAY BY THE MARS *GLOBAL SURVEYOR* PROVIDE WIDE-ANGLE CAMERAS A GLOBAL "SNAPSHOT" OF WEATHER PATTERNS ACROSS MARS. HERE, BLUISH-WHITE ICE CLOUDS HANG ABOVE THE THARSIS VOLCANOES.

PHOTO: NASA Jet Propulsion Laboratory/Malin Space Science Systems

READING SELECTION

EXTENDING YOUR KNOWLEDGE

In 2003, the United States landed two more rovers on Mars. One was named *Spirit*, the other *Opportunity*. These two robots landed at different locations on Mars. *Spirit* landed within the Gusev Crater. Three weeks later *Opportunity* landed on Mars at Terra Meridiani. These rovers have found evidence of water on Mars, taken numerous photographs of rocks and other surface features, and sent back information about weather conditions. Designed to work for months, the rovers worked for years. *Spirit* continued to transmit data for months even after it got permanently stuck in soft Martian soil. The two rovers turned out to be very hardy explorers.

Two years later, in 2006, the *Reconnaissance* orbiter arrived at Mars. It went into orbit around the planet and has sent back detailed views of the surface of Mars and of clouds and other weather phenomena. This information will be valuable for future missions to the red planet.

The spacecraft *Phoenix* arrived in 2008 and landed near the north pole of Mars. It explored the Martian ice caps and identified the presence of water in the soil. The presence of water is essential for life as we know it. Its mission ended in November 2008, when it lost solar power as the Martian winter came on.

▶ *PHOENIX* MARS LANDER
TOUCHING DOWN NEAR
THE NORTH POLE OF MARS

PHOTO: NASA Jet Propulsion Laboratory—Caltech/University of Arizona

READING SELECTION

EXTENDING YOUR KNOWLEDGE

We will keep on exploring the red planet, hunting for signs of life. Ironically, the strongest evidence we have found of past life on Mars was actually found on Earth. The Allan Hills meteorite, originally part of Mars, was found in Antarctica by a team of researchers in 1984. Since then, astronomers, biologists, and archaeologists have argued about whether or not it contains fossils of Martian life, billions of years old. The pendulum has swung most recently in favor of the fossils, but a discovery of fossils on Mars would clinch the case. ■

DISCUSSION QUESTIONS

1. Compare missions to Venus with missions to Mars. What differences in conditions have the spacecraft faced, and what have they been able to find out about the respective planets?

2. If we tried to live on Mars, what would be some of the challenges we'd encounter? How might we solve them?

NAME THIS ROVING ROBOT!

Imagine what it would be like to name the rover for the Mars Pathfinder mission. Valerie Ambrose, a 12-year-old student from Bridgeport, Connecticut, didn't have to imagine. In 1995, she won a contest held by the Planetary Society, a nonprofit organization dedicated to the exploration of the solar system.

Kids between the ages of 5 and 18 could enter the contest. The Society said that the rover had to be named for a heroine from mythology, fiction, or history (no longer living). Entrants had to submit a fully researched, 300-word essay explaining their choice of a name for the rover.

Entries came from around the world. Valerie suggested that the rover be named after Sojourner Truth, the African-American activist who wanted slavery abolished and who promoted women's rights. Sojourner Truth lived during the Civil War era, and she traveled around the United States speaking for the rights of all people to be free and for women to fully participate in society.

Finally, the day to name the winner arrived. Valerie won! NASA chose the name *Sojourner* for its Mars *Pathfinder* rover. The name honors Sojourner Truth. It is also appropriate because "sojourner" means "traveler."

Another student's entry was also a winner. Second-place winner Deepti Rohatgi suggested Marie Curie, after the Polish-born chemist who won the Nobel Prize in 1911 for her discovery of the elements radium and polonium. NASA used *Marie Curie* as the name for another Mars rover.

It's exciting to think that students like you named these famous rovers. Who knows, maybe one day you may name a rover, a comet, an asteroid, or even the next planet found in space!

PLANETARY FACTS

MARS

MARS: QUICK FACTS*

▸ **Diameter**
6779 km

▸ **Average distance from the Sun**
227,943,824 km

▸ **Mass**
64×10^{22} kg

▸ **Surface gravity (Earth = 1)**
0.38

▸ **Average temprature**
-65°C

▸ **Length of sidereal day**
1.03 Earth days

▸ **Length of year**
686.98 Earth days

▸ **Number of observed moons**
2

COMPOSITION

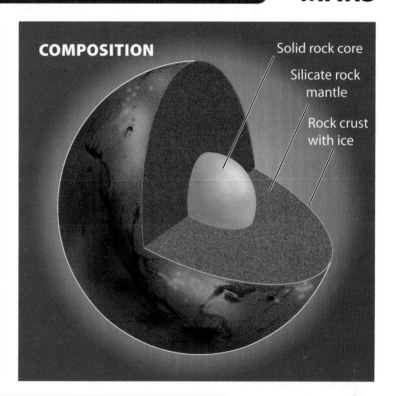

Solid rock core

Silicate rock mantle

Rock crust with ice

RELATIVE SIZE

DID YOU KNOW?

▸ Mars was named after the Roman god of war because the red color of Mars looks like spilled blood.

▸ Olympus Mons, a volcano on Mars, is 24 kilometers high—more than twice as high as the tallest volcano on Earth, and almost three times as high as Mount Everest.

ATMOSPHERE

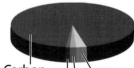

Carbon dioxide (95.3%)

Nitrogen (2.7%)

Argon (1.6%)

Oxygen, water vapor, and others (minor amounts)

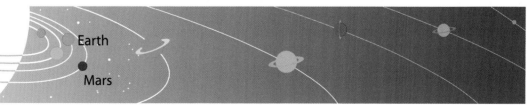

Earth

Mars

* Source: Data from NASA as of 2011

GRAVITY AND ORBITAL MOTION

INTRODUCTION

How does a manmade satellite get into orbit? A satellite is launched by a rocket to a height (or altitude) at which Earth's gravitational force keeps the satellite in orbit around Earth. A satellite, like the Moon, must travel at just the right speed to stay in Earth's orbit. If the satellite moves too slowly, gravity might pull it back down to Earth. If the satellite moves too fast, it might escape Earth's gravitational pull and zoom out into space.

In Lesson 3, you investigated the effects of surface gravity on weight. In this lesson, you will conduct four inquiries that focus on gravity and its effects on the orbits of moons and planets. What part does gravity play in keeping the planets in orbit around the Sun? How do the moons stay in orbit around each planet? In this lesson, you will investigate these and other questions. You will also read to learn more about missions to Saturn, Uranus, and Neptune.

▶ **WHAT KEEPS A SATELLITE ORBITING IN SPACE? GRAVITY! GRAVITY ALSO MAKES PLANETS LIKE EARTH ORBIT THE SUN.**

PHOTO: NASA Jet Propulsion Laboratory—Caltech

OBJECTIVES FOR THIS LESSON

- Analyze patterns in planetary motion.

- Observe the motion of a marble when acted upon by different forces.

- Investigate the effect of a pulling force on the orbital period of a sphere.

- Relate the observed behavior of a marble and sphere to the motion of moons and planets.

- Summarize, organize, and compare information about Saturn, Uranus, and Neptune.

▶ **MATERIALS FOR LESSON 7**

For you

1	working copy of Student Sheet 2.3c: Solar System Chart
1	pair of goggles

For the class

1	Inquiry Card Set
1	plastic box from Lesson 5 (filled with sand, flour, and cocoa)
2	metric rulers, 30 cm (12 in)
3	marbles
1	metric measuring tape
1	plastic box (empty)
1	metal canning jar ring
1	sheet of white paper
1	Planetary Motion Model™
4	plastic boxes or boxes of the same height
1	yellow balloon, filled with water
1	pre-assembled Moon Orbiter™
25	large steel washers
1	student timer

GETTING STARTED

1 Review the Introduction section of the software *Explore the Planets* with your class.

2 Use what you learned from *Explore the Planets* to make general observations about the planets' motion around the Sun. Record your ideas in your science notebook if instructed to do so by your teacher. Discuss your ideas with the class. ✍

PROCEDURE FOR THE CIRCUIT OF INQUIRIES

1 Review with your teacher the procedure for completing the circuit of four inquiries in this lesson.

2 Review the Procedure and Safety Tips for each inquiry before you begin the circuit.

3 To record your observations, divide your science notebook page into quadrants. Label the quadrants 7.1, 7.2, 7.3, and 7.4. ✍

4 Your teacher will assign your group to a station where you will begin the circuit of inquiries.

5 Complete all four inquiries in the order given to you by your teacher. Remember to return all of your equipment and the Inquiry Card to the resealable plastic bag before moving on to the next station.

▶ **WHAT FORCES KEEP THE MOON IN ORBIT AROUND EARTH?**

PHOTO: NASA Jet Propulsion Laboratory

GRAVITY'S EFFECT ON OBJECTS IN MOTION

SAFETY TIPS

Wear safety goggles at all times.

Work in a well-ventilated area to minimize the level of dust in the air.

PROCEDURE

1. Hold the marble 40 cm above the plastic box. With the marble in your hand, decide what two forces are acting on the marble. Are the forces balanced (both pulling equally) or unbalanced (one is pulling more than the other)? Discuss your ideas with your group.

2. What will happen if you release the marble from your hand into the box? Discuss your predictions with your group.

3. Let go of the marble. Discuss your observations of the marble's motion with your group. (Do not be concerned about the crater that the marble makes. The sand and flour keep the marble from moving once it lands in the box.) Compare your observations to your predictions.

▶ **ROLL THE MARBLE DOWN THE RULER INTO THE PLASTIC BOX.**
FIGURE **7.1**

4. Repeat Procedure Steps 1–3. Does the marble move the same way each time? Discuss your observations and record them in quadrant 7.1 in your science notebook.

5. Use the ruler as a ramp to gently roll the marble into the plastic box, as shown in Figure 7.1. Keep the ruler nearly flat. Discuss your observations. How did the marble move once it left the ruler?

Inquiry 7.1 continued

6 Experiment by rolling the marble down the ruler at different speeds. Keep the ruler nearly flat. How does the marble move each time it leaves the ruler? If possible, measure the distance that your marble travels each time. Record your observations.

7 Record your answers to these questions:

A. What pulling force acts on the marble at all times?

B. When you rolled the marble slowly, how did it move once it left the ruler?

C. How does the forward speed of the marble affect the motion of the marble once it leaves the ruler?

D. All planets that orbit the Sun are traveling forward due to inertia and falling toward the Sun due to gravity. Describe the path of something that has forward motion (like your marble) but is also being pulled down by gravity.

8 Clean up. Return all materials to their original condition.

INQUIRY 7.2

TESTING BALANCED AND UNBALANCED FORCES

SAFETY TIP
Wear safety goggles at all times.

PROCEDURE

1 Place the white paper in the bottom of the plastic box. Put the metal ring on top of the paper with the lip up, as if you were putting the metal ring on a jar.

2 Trace an outline of the ring onto the paper. Remove the metal ring from the paper. Mark four points at equal distances around the circle. Number the marks 1 to 4 going clockwise, as shown in Figure 7.2.

3 Place the metal ring on the circle, again with the lip up. Place the marble inside the metal ring. Without moving the metal ring, describe the motion of the marble. Record your observations in quadrant 7.2 in your science notebook. ☑

▶ MARK THE OUTLINE OF THE CIRCLE WITH
1, 2, 3, AND 4 AT QUARTER INTERVALS.
FIGURE **7.2**

Inquiry 7.2 continued

4 Use the ring to move the marble in circles. Keep the ring on the paper at all times. Record your observations. Discuss with your group how the ring creates a force (called an "unbalanced force") that influences the marble's motion.

5 Make a prediction about what will happen if you lift the ring (remove the unbalanced force).

6 Move the marble in circles again, then lift the ring. What happens? Write a description of the motion of the marble without the unbalanced force of the ring. Try this several times. Record your observations in both words and pictures. Use your numbered markings to pinpoint the motion of the marble each time.

7 Record your answers to these questions:

A. Describe the motion of the marble when an unbalanced force (the metal ring) influences it.

B. Describe the motion of the marble when the unbalanced force is removed.

C. Suppose you lifted the ring when the clockwise orbiting marble was at the "1." Draw the path the marble would take.

D. Suppose you lifted the ring when the clockwise orbiting marble was at the "4." Draw the path the marble would take.

E. Like the marble, the planets move forward due to inertia and inward due to an unbalanced force. Together, these forces cause the planets' paths to curve. What is the unbalanced force that keeps the planets in orbit? What would happen to the planets without this unbalanced force?

8 Clean up. Return all materials to their original condition.

OBSERVING PLANETARY MOTION

PROCEDURE

1 Check the setup of the Planetary Motion Model™. The lip of the hoop should be facing up to prevent the marble from falling off the latex sheet, as shown in Figure 7.3. Allow any extra sheeting to hang down under the hoop. Make sure the hoop rests on the edges of the boxes so they do not interfere with the marble once it is on the sheet.

SAFETY TIPS

Wear safety goggles at all times.

Be careful working with the balloon. It can be a choking hazard.

The latex in the rubber sheeting may cause either an immediate or delayed allergic reaction in certain sensitive individuals.

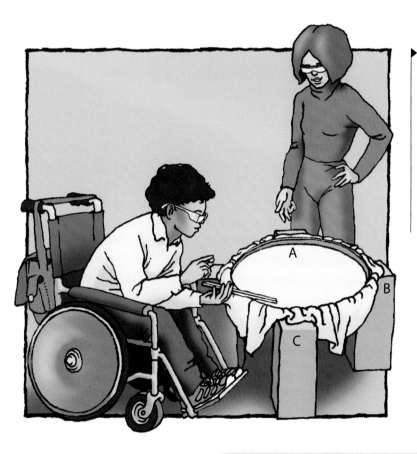

▶ THE PLANETARY MOTION MODEL™ SHOULD BE SET UP AS SHOWN. (A) FACE THE LIP OF THE HOOP UP. (B) HANG THE EXTRA SHEETING UNDER THE HOOP. (C) PLACE THE HOOP ON THE EDGE OF EACH BOX.
FIGURE **7.3**

Inquiry 7.3 continued

2 You will use your ruler as a ramp to roll the marble onto the latex sheet. Before you do, make a prediction about the path the marble will take on the sheet. Discuss your predictions with your group.

3 Hold the ruler as shown in Figure 7.3 so that it faces the edge of the hoop. Roll the marble onto the flat sheet. Observe the marble. Repeat this several times. Discuss your observations with your group. Record your results in quadrant 7.3 in your science notebook. ✎

4 Place the water-filled balloon in the center of the sheet. Let go of the balloon. Discuss what the balloon does to the sheet. Then roll the marble onto the sheet toward the edge of the hoop. Watch the balloon and marble carefully. What do you observe about the motion of the marble? What do you observe about the behavior of the balloon? Discuss and record your observations with your group.

5 Now push down on the balloon as shown in Figure 7.4. Keep a constant pressure on the balloon.

▶ PRESS DOWN CONTINUOUSLY ON THE WATER-FILLED BALLOON.
FIGURE **7.4**

6 Predict how the marble will move now that the center of the sheet has more mass. Have one of your partners roll the marble onto the sheet as you keep pressure on the balloon. Discuss your observations. Record your observations in words and pictures.

7 Test the motion of the marble several times and observe its motion carefully. Let everyone take a turn. How does the motion of the marble change as it nears the balloon?

8 Now wobble the balloon very slightly as the marble orbits it. What happens? Try to use a gentle wobble on the balloon to keep the marble in motion. Discuss your observations. Then let go of the balloon. Does the balloon wobble on its own as the marble orbits it?

9 Record your answers to these questions:

A. Describe how the marble moved when the mass in the center (the balloon) was not present.

B. Describe how the marble moved when the mass in the center was present.

C. As the distance between the balloon and marble decreased, what happened to the marble's speed?

D. Based on your observations, which planet do you think would have the fastest orbital speed? What evidence do you have to support your answer?

E. What force keeps the planets in their orbital paths around the Sun?

F. Read "Stars Wobble." Why does a star's "wobble" indicate that a planet is nearby?

10 Clean up. Return all materials to their original condition.

BUILDING YOUR UNDERSTANDING

STARS WOBBLE

There are many stars like our Sun. Some of these other stars may also have planets that orbit them. In 2008, the Hubble Space Telescope took the first picture of an exoplanet (a planet orbiting another star). The exoplanet is three times more massive than Jupiter. But most planets around other stars cannot be seen directly. So how do astronomers know that there are planets around other stars?

One way is to observe a star's motion. When a planet orbits a star, it makes the star wobble. The wobble will make the star move toward us and away from us. Astronomers can detect this motion using spectroscopes attached to their telescopes.

It all begins with gravity. The law of gravity tells us that all objects pull on each other. The Sun pulls on the planets, but it also means that the planets pull on the Sun. (And moons and planets tug at each other.) An orbiting planet exerts a gravitational force that makes the star wobble in a tiny circular or oval path. The star's wobbly path mirrors in miniature the planet's orbit. It's like two twirling dancers tugging each other in circles.

From the size of the wobble, astronomers can figure out how big, how massive, and how far away the planet is from the star. This method has been very successful for finding exoplanets. In 1995, astronomers confirmed the discovery of the first exoplanet. By 2007, more than 200 planets had been found. How many exoplanets have astronomers found now? See if you can find out. ■

INVESTIGATING THE EFFECT OF PLANETARY MASS ON A MOON'S ORBIT

SAFETY TIPS

Wear safety goggles at all times.

Do not swing the Moon Orbiter at other students. Make sure that other students are not nearby when you swing the white sphere.

Always swing the Moon Orbiter above your head.

PROCEDURE

1 Examine the Moon Orbiter™. Discuss with your group how you think the Moon Orbiter might work.

2 Move to an area in the classroom where no other groups are working. Check to see that all nylon knots are secured to the large white sphere.

3 Hold the narrow plastic tubing of the Moon Orbiter in your hand like a handle. Practice holding the Moon Orbiter over your head and moving your hand in circles to get the white sphere to orbit your hand. Use a steady and regular motion. When the sphere is in full orbit, the bottom of the tube should nearly touch the cylinder. Let everyone in your group try to swing the Moon Orbiter. Remember, when the sphere is in full orbit, the tube should nearly touch the cylinder.

▶ **SWING THE WHITE SPHERE IN A CIRCLE ABOVE YOUR HEAD.**
FIGURE **7.5**

4 Increase the mass of the Moon Orbiter by adding five washers to the cylinder. Move your hand in circles over your head to get the white sphere to orbit your hand, as shown in Figure 7.5. Describe how fast the sphere has to move to stay in orbit around your hand with a mass of five washers pulling on it. (If possible, calculate its orbital period—the time it takes the sphere to orbit your hand. For example, count the number of seconds it takes the sphere to orbit your hand 10 times. To get the orbital period, divide the number of seconds by 10.) Record your observations and data in quadrant 7.4 in your science notebook. ☞

5 Predict what will happen if you increase the mass of the Moon Orbiter's cylinder to 25 washers.

6 Fill the cylinder of the Moon Orbiter with 25 washers. Repeat Procedure Step 4 and discuss your observations. Describe how fast the sphere has to move to stay in orbit around your hand with 25 washers pulling on it. (Try calculating the sphere's orbital period.) Record your observations.

Inquiry 7.4 continued

7 Record your answers to these questions:

A. How does the mass of the cylinder affect how fast or slow the sphere orbits your hand?

B. Examine Table 7.1. Compare the mass of Jupiter with the mass of Earth. Which planet has more mass?

C. Examine Table 7.1. Compare Jupiter's moon Io with Earth's moon. How are they alike? How are they different?

D. Compare Io and the Moon. Which planetary satellite travels faster (has a greater orbital speed)? Given your results from the inquiry, why do you think this is?

E. Orbital period is the time it takes a revolving object to orbit a central object. Which planetary satellite has a shorter orbital period? What is the relationship between orbital speed and orbital period?

F. In Lesson 3, you learned the approximate mass of each planet. How do you think scientists determine the mass of the planets?

TABLE 7.1 PLANETARY MASS VERSUS MOON'S ORBITAL PERIOD

SOLAR SYSTEM BODY	MASS (kg)	DIAMETER (km)	DISTANCE FROM PLANET (km)	ORBITAL SPEED (km/sec)	ORBITAL PERIOD (DAYS)
JUPITER	$189{,}813 \times 10^{22}$	139,822			
EARTH	597×10^{22}	12,742			
IO	9×10^{22}	3643	421,800	17	2
MOON	7×10^{22}	3475	384,400	1	27

Source: http://solarsystem.nasa.gov

8 Clean up. Return all materials to their original condition.

REFLECTING
ON WHAT
YOU'VE DONE

1 Share your answers to the inquiry questions with the class.

2 Read "Heavy Thoughts" on pages 136–141. Be prepared to discuss the questions at the end of the reading selection.

3 With your class, return to the Question F (What is gravity? Where is gravity strongest? Where is it weakest? Why?) and Question G (What keeps the planets in orbit around the Sun?) folders from Lesson 1. Is there anything you would now change or add? Discuss your ideas with the class.

4 Return to your list of ideas about gravity from Lesson 3. What new information about gravity do you want to add to your list?

Heavy Thoughts

Do you ever wonder why when you jump up, you always come back down? Or do you ever wonder why the Moon keeps circling around Earth rather than drifting off into space? Throughout history, people have wondered about these things. Now we know that a force called gravity is responsible. Gravity governs the movements of everything on Earth, and all the objects in the sky.

NEWTON'S APPLE

According to a well-known story, a brilliant 23-year-old English scientist named Isaac Newton was sitting under an apple tree one afternoon in 1666, when he saw an apple fall to the ground. Newton began thinking about the force that pulled the apple from the tree.

After careful experimentation, Newton concluded that there must be a single, invisible force responsible for these motions. The force we know as gravity is like the one you can feel when you place a magnet near a metal object (although gravity is not as strong as electromagnetic forces). Gravity makes apples fall from trees and it holds planets and moons in their orbits.

▶ A FAMOUS STORY SAYS THAT ISAAC NEWTON BEGAN THINKING ABOUT GRAVITY WHEN HE SAW AN APPLE FALL FROM A TREE.

WHAT IS AN UNBALANCED FORCE?

If two individual forces are of equal magnitude (size) and opposite direction, then the forces are balanced. Think of the marble you held in your hand during Inquiry 7.1. One force—Earth's gravitational pull—exerts a downward force on the marble. The other force—your hand—pulls upward on the marble. The forces acting on the marble are balanced; as a result, the marble is at rest. Its motion does not change; it does not slow down or speed up or change direction. But if the two forces are not balanced, the marble will change its motion. It will change its speed or direction or both. For example, when you let go of the marble, the unbalanced force of gravity disturbs the marble's motion and the marble falls into the box. As it falls, its speed constantly increases. Unbalanced forces cause objects to accelerate (change their speed or direction).

▶ GRAVITY KEEPS THE EIGHT PLANETS AND THEIR MOONS, DWARF PLANETS, AND THOUSANDS OF COMETS AND ASTEROIDS IN ORBIT AROUND THE SUN.

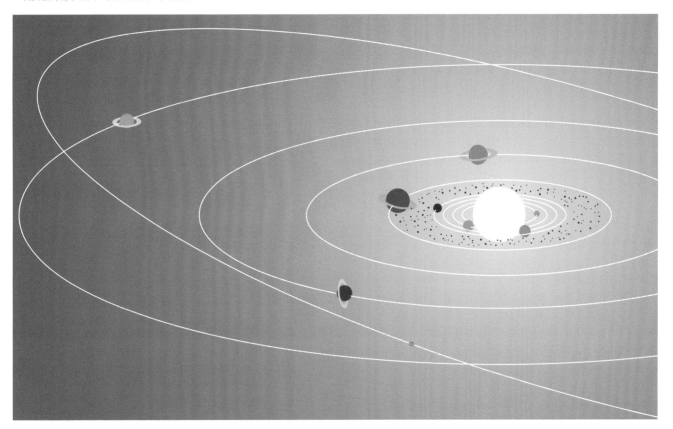

READING SELECTION

NEWTON'S LAW OF UNIVERSAL GRAVITATION

Newton's Law of Universal Gravitation states that any two objects in the universe have gravity and will attract each other. Just how much those objects attract each other depends on two things—the mass of each object, and the distance between the objects.

The more mass a star (like our Sun) has, and the closer a planet is to that star, the greater the star's ability to hold the planet in its orbit. Mercury, for instance is very close to the Sun, and it's unlikely to drift away. Also, the more massive a planetary object, the better it is at pulling faraway objects into orbit. Jupiter, a planet with a lot of mass, has 62 known moons in orbit around it.

NEWTON'S LAWS OF MOTION

Newton described three famous laws of motion that help us understand how forces, like gravity, affect the motion of bodies.

The first of Newton's three laws of motion is called the Law of Inertia. It says that a body in motion tends to travel at a constant speed in a straight line unless it is disturbed by an unbalanced force (a force that isn't canceled out by another force). The Law of Inertia explains why you don't keep rising when you jump up in the air. The unbalanced force of gravity disturbs your motion and pulls you back down. The Law of Inertia governs the motion of the planets and moons. If they weren't affected by gravity, they would leave the solar system immediately, traveling in straight lines like a key you've been twirling when you let its cord fly off your finger. Gravity holds all the planets in orbit around the Sun, and each planet's gravity captures and holds its moon(s) in orbit.

Newton's Second Law of Motion tells how unbalanced forces affect the motion of objects. It says that unbalanced forces will change the motion of an object.

MUTUAL ATTRACTION

An object with a large amount of mass can exert a huge gravitational pull even on objects that are quite distant and massive. The Sun's gravitational pull is so enormous that it easily hangs onto Jupiter, which has a mass two-and-a-half times as much as all the other planets combined. The Sun also exerts a gravitational hold over dwarf planets like Pluto. But, tiny Pluto also exerts a small gravitational pull on the Sun, even though they are more than 4.5 billion kilometers apart!

ACTION AND REACTION

Newton's Third Law of Motion tells us that objects exert equal and opposite forces on one another. The Sun pulls on the planets and the planets pull on the Sun. The planets pull on the moons that orbit them and the moons pull on the planets. The same thing happens between you and Earth. If you jump up, Earth's gravity pulls you back down and your gravity also pulls Earth toward you.

ORBITAL VELOCITY

The farther a planet is from the Sun, the more slowly it travels in its orbit. The closer a planet is to the Sun, the faster it travels in its orbit. Mercury, the planet closest to the Sun, travels at about 48 kilometers (30 miles) per second. But Neptune, the planet farthest from the Sun, is quite a different story. Look at Table 7.2: Orbital Velocity of Planets on page 141 and compare the orbital velocity of the planets. Do you notice patterns in the data? If so, what are they?

The attraction between two objects decreases as the distance between them increases. The

▶ THE GREAT MASS OF JUPITER HELPS HOLD
ITS MANY MOONS IN ORBIT AROUND THE
GIANT PLANET.

PHOTO: NASA Jet Propulsion Laboratory

READING SELECTION

EXTENDING YOUR KNOWLEDGE

HOW MATTER AFFECTS SPACE

Is gravity a force, or is it something else? About 250 years after Newton, another genius started thinking about gravity: Albert Einstein. Einstein's theories changed the way we think about the universe.

Einstein came to believe that gravity isn't really a force, but simply the way that matter affects space. According to Einstein, wherever there's a chunk of matter—an apple, a person, a planet, or a star—the matter curves the space around it. The bigger the matter, the more that space is curved. And when space is curved, anything traveling through that space must follow those curves.

According to Einstein, the planets are caught in the curved space around the Sun. Our moon is caught in the curved space around Earth. If you were far enough away from the gravitational force of Earth or the Sun, small objects would become caught in the curved space around you!

MODELING CURVED SPACE

Einstein believed that the more massive the object, the more it curved the space around it. Think back to the inquiry in which you placed a water-filled balloon in the center of a rubber sheet. The balloon curved the rubber sheet around it. A marble placed on the sheet rolled toward the balloon, but not in a straight line. Instead, the marble followed the curves of the sheet and "orbited" the water-filled balloon in the center. The closer the marble got to the balloon in the center, the faster the marble rolled. Something similar happens with stars such as the Sun. Space curves around the star's mass and helps to keep other objects, such as planets, "rolling" around them.

▶ MATTER CURVES THE SPACE AROUND IT, JUST LIKE A HEAVY BALLOON CURVES THE SURFACE OF A RUBBER SHEET WHEN THE BALLOON RESTS ON IT.

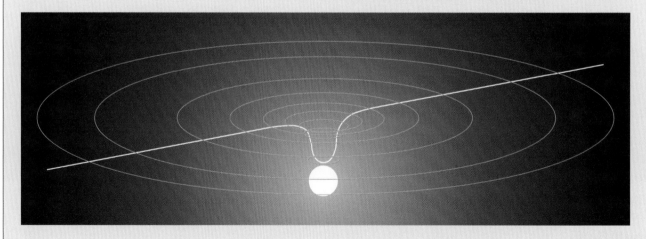

Sun's pull on distant Neptune is much less than its pull on nearby Mercury. As a result, Neptune orbits the Sun at a much slower speed.

Newton and other scientists made important discoveries that describe how gravity works not just on Earth, but throughout the universe. Newton's laws of motion and universal gravitation enable us to predict how planets, moons, and other bodies in our solar system move. That's made it possible to send probes and spacecraft to planets, moons, comets, and asteroids. After all, if you can't predict where Mars will be when your spacecraft reaches its orbit, your chances of landing on the planet are not so good. ■

DISCUSSION QUESTIONS

1. What would happen to the planets if there were no gravitational influences from the Sun?

2. In baseball, why does a curve ball curve? Explain using Newton's laws of motion.

TABLE 7.2 ORBITAL VELOCITY OF PLANETS

PLANET	ORBITAL VELOCITY (km/s)	APPROXIMATE DISTANCE FROM SUN (km)
MERCURY	47	57,900,000
VENUS	35	108,200,000
EARTH	30	149,600,000
MARS	24	227,900,000
JUPITER	13	778,300,000
SATURN	10	1,426,700,000
URANUS	7	2,870,700,000
NEPTUNE	5	4,498,400,000

Source: http://solarsystem.nasa.gov

Mission: SATURN

VOYAGERS 1 AND 2

The twin spacecraft, *Voyagers 1* and *2*, left Earth in the summer of 1977. Three years later in 1980, after its flyby of Jupiter, *Voyager 1* flew past Saturn. In 1981, *Voyager 2* flew past Saturn.

The *Voyager* probes gave us a new, sharper view of Saturn, its rings, and its moons. The rings are like a brilliant, vivid necklace with 10,000 strands, and they proved to be more beautiful and strange than once thought.

Evidence indicates that Saturn's rings formed from large moons that were shattered by comets and meteoroids that struck them, exploding into huge clouds of debris. The resulting ice and rock fragments—some as small as a speck of sand and others as large as houses—went on orbiting Saturn, gathering in a broad plane around the planet. The rings themselves are very thin, but together they are 171,000 kilometers in width! (That's more than 10 times Earth's diameter.)

The irregular shapes of Saturn's eight smallest moons indicate that they, too, are fragments of larger bodies. Two of these small moons—Prometheus and Pandora—are located in one of Saturn's many rings.

Voyager 2 showed something even more dramatic: a gravitational tug of war among Saturn, its 60+ moons and moonlets, and the ring fragments. This struggle has caused variations in the thickness of the rings. Some particles are even rising above the ring band as if they might escape.

READING SELECTION

EXTENDING YOUR KNOWLEDGE

THE CASSINI MISSION

In October 1997, NASA launched a probe on a new mission to explore Saturn. The probe was named *Cassini*, in honor of the seventeenth century Italian astronomer who observed Saturn and discovered a gap in Saturn's ring system, and its launch generated worldwide excitement. In order to save fuel, the *Cassini* probe used gravity assists from Venus, Earth, and Jupiter along its journey to the distant ringed planet. *Cassini* arrived at Saturn in 2004 and went into an elliptical orbit around the planet.

For four years, *Cassini* sent back fascinating images of Saturn's rings and moons. Saturn's rings look smooth and orderly as seen from Earth, but the *Cassini* images showed that the ring system around Saturn is complex. Small moons and ring particles collide within the ring system and waves form within the rings as well.

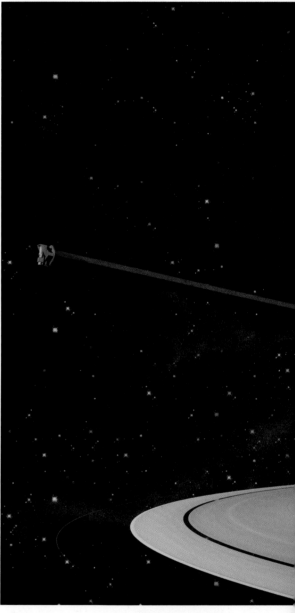

▶ **THIS IMAGE OF IAPETUS FROM THE CASSINI MISSION SHOWS THE CONTRAST BETWEEN THE WHITE AND BLACK SIDES OF THE MOON.**

PHOTO: NASA Jet Propulsion Laboratory/Space Science Institute

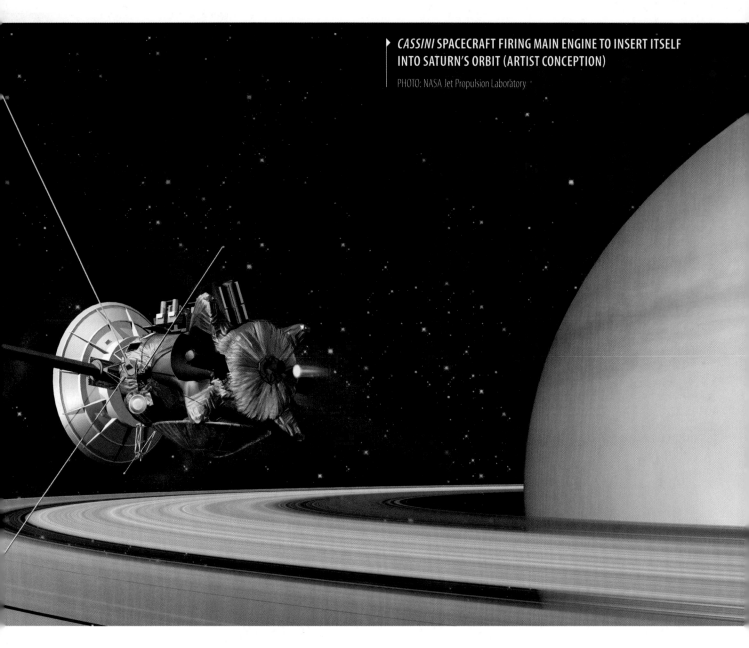

Cassini also looked closely at the cloud patterns in Saturn's atmosphere to learn about weather on the planet. *Cassini* found that Saturn has rain and snow and storms with lightning.

Images of Saturn's moons revealed fantastic features, such as ice plumes that contain carbon-backbone chemicals like those found on Earth, and rings around some moons, as well as puzzling surface features like an unusual bulge in one moon's dark-side shadow. *Cassini* found that the moon, Iapetus, has a white side, a black side, and a mountainous bulge around its equator. *Cassini* also discovered previously unseen moons orbiting Saturn; more than 60 moons have now been counted.

READING SELECTION
EXTENDING YOUR KNOWLEDGE

▶ WORKERS GETTING THE *HUYGENS* PROBE READY FOR ITS MISSION.

PHOTO: NASA Kennedy Space Center

Cassini carried another probe inside itself. The probe was named *Huygens*, in honor of Christian Huygens, who discovered Saturn's largest moon, Titan, in 1655. The *Huygens* probe was built by the European Space Agency and its mission was to descend through the atmosphere of Titan and send back data. It successfully completed that mission on January 14, 2005. *Huygens* survived the trip through the atmosphere and landed on the surface of Titan. *Huygens* found huge methane lakes, regions of wind-driven hydrocarbon sand dunes, and ice boulders on Titan.

The original Cassini mission ended in 2008, but it not only answered questions about Saturn and its rings and moons, it generated many new questions—much to the excitement of astronomers! The Cassini mission was extended and renamed the Saturn Equinox mission. The new mission has given us more time to learn more about this fascinating system. ■

DISCUSSION QUESTIONS

1. Compare and contrast the characteristics of Saturn with those of other gaseous giants.

2. Given what you know about requirements for life, would it be possible for life to exist on Saturn? Is it likely? Explain your ideas.

PLANETARY FACTS SATURN

SATURN: QUICK FACTS*

▶ **Diameter**
116,464 km

▶ **Average distance from the Sun**
1,426,666,422 km

▶ **Mass**
$56,832 \times 10^{22}$ kg

▶ **Surface gravity (Earth = 1)**
1.06

▶ **Average temperature**
-140°C

▶ **Length of sidereal day**
10.66 hours

▶ **Length of year**
29.45 Earth years

▶ **Number of observed moons**
62

COMPOSITION

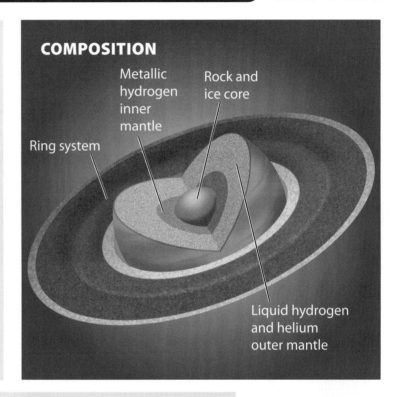

Metallic hydrogen inner mantle

Rock and ice core

Ring system

Liquid hydrogen and helium outer mantle

RELATIVE SIZE

DID YOU KNOW?

▶ If there were an ocean big enough to plop Saturn into, the planet would float—just like an iceberg does on planet Earth! That's because of Saturn's low density. Saturn is the only planet that is lighter than the same volume of water.

▶ Saturn's winds reach 1800 km/hour. (The strongest tornadoes on Earth have wind speeds of only 350 km /hour.)

ATMOSPHERE

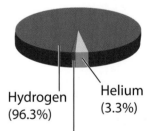

Hydrogen (96.3%)

Helium (3.3%)

Water ice, methane, ammonia, and other compounds (traces)

Earth

Saturn

* Source: Data from NASA as of 2011

Mission: On to Uranus and Neptune

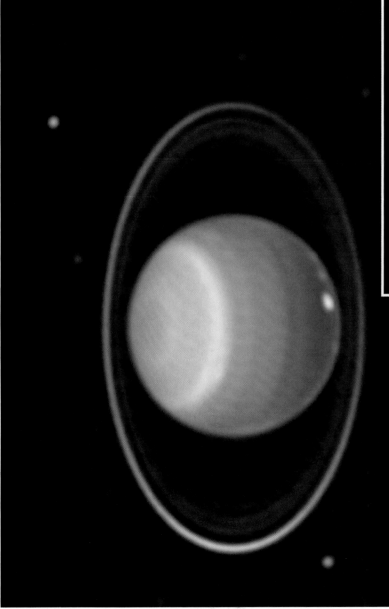

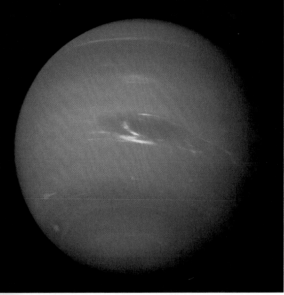

▶ **(ABOVE) NEPTUNE IS TRUE BLUE.**

PHOTO: NASA Jet Propulsion Laboratory

▶ **(LEFT) THE RINGS OF URANUS. NOTICE THAT URANUS IS TILTED 98 DEGREES ON ITS AXIS.**

PHOTO: NASA Jet Propulsion Laboratory/Space Telescope Science Institute

Beginning in 1977, the Voyager mission became quite a scientific odyssey! After flying past Saturn, *Voyager 1* left the plane of the solar system, as planned. (If you imagine the planets laid out on a flat disc around the Sun, you can see what's meant by "the plane of the solar system.") *Voyager 2* was supposed to take the same course. But the

READING SELECTION
EXTENDING YOUR KNOWLEDGE

spacecraft was performing so well, scientists and engineers on Earth decided to send it on to Uranus and Neptune for a closer look. Saturn, Neptune, and Pluto were lined up near each other in their orbits, and a similar alignment of the outer planets would not occur again until the year 2157. Never before had we had a close-up view of the outer solar system.

Saturn's huge gravitational field pulled *Voyager 2*'s path toward Saturn, then flung the spacecraft toward Uranus. A similar boost from giant Uranus would send *Voyager 2* to Neptune. This maneuver, called a gravity assist, took decades off *Voyager*'s flying time.

Unfortunately, the grand tour of the solar system conducted by *Voyager* couldn't include Pluto because Pluto's orbit took it far from the spacecraft's path. Still, the remarkable journey of *Voyager 2* yielded many insights.

▶ THIS VIEW OF URANUS WAS ACQUIRED BY *VOYAGER 2* IN JANUARY 1986. THE GREENISH COLOR OF ITS ATMOSPHERE IS DUE TO METHANE AND SMOG. METHANE ABSORBS RED LIGHT AND REFLECTS BLUE-GREEN LIGHT.

PHOTO: NASA Jet Propulsion Laboratory

URANUS

After its push from Saturn's gravitational field, *Voyager 2* arrived at Uranus in 1986, and on arrival found 10 moons never seen by Earth's telescopes. That brought the count of Uranus's moons to 20. Scientists had believed that there might have been several more tiny moons within the rings, and they were right!

The *Voyager* cameras also found that Uranus had more rings than had been seen from Earth, and showed that rings of fine dust surrounded the planet's nine major rings.

According to data, Uranus's rings probably formed after the planet itself. Like Saturn's rings, Uranus's appeared to be the remnants of a moon destroyed in a collision.

Voyager 2 made another major discovery at Uranus. It turns out that the planet has a magnetic field as strong as Earth's. The cause of this field isn't clear, but it's shaped like a long corkscrew. And, according to *Voyager*'s measurements, it reaches 10 million kilometers (more than 6 million miles) behind the planet! How far is that? Earth's magnetic field extends just tens of thousands of kilometers, so this is a couple of orders of magnitude greater.

NEPTUNE

Thanks to the gravity assist from Uranus, *Voyager* reached Neptune just three years later, in 1989. Until that time, scientists had believed that the planet had arcs, or partial rings. But *Voyager* showed that Neptune has complete rings with bright clumps; it was the clumps that had been noticed from Earth. *Voyager* also discovered six new moons.

THIS 1998 IMAGE SHOWS THE GREAT DARK SPOT ON NEPTUNE, A STORM RAGING IN THE NORTHERN HEMISPHERE. THE WHITE SLASH BELOW IT IS A GROUP OF CLOUDS, SIMILAR IN APPEARANCE TO CIRRUS CLOUDS ON EARTH, BUT MADE FROM CRYSTALS OF FROZEN METHANE RATHER THAN ICE. THESE STORMS COME AND GO; A PREVIOUS GREAT DARK SPOT IN THE SOUTHERN HEMISPHERE DISAPPEARED IN 1994.

PHOTO: NASA Jet Propulsion Laboratory

READING SELECTION

Voyager carries not just scientific equipment, but a message for any intelligent life forms it might meet. On a golden record, the kind of record made for old-fashioned record players, it carries the following symbols, as well as several kinds of recordings, all aimed at telling the alien about Earth, Earthlings, and the space mission. Can you guess what the symbols mean?

The record carried music from around the world; Earth sounds like wind, whales, and a kiss; the brainwaves of the wife of astronomer Carl Sagan; and a printed message from President Jimmy Carter. It said: "This is a present from a small, distant world, a token of our sounds, our science, our images, our music, our thoughts, and our feelings. We are attempting to survive our time so we may live into yours."

Voyager flew within 5000 kilometers (3107 miles) of Neptune's long, bright clouds, which resemble cirrus clouds on Earth. Instruments aboard *Voyager* measured winds up to 2000 kilometers (1243 miles) an hour—the strongest winds on any planet. Tornado-strength winds on Earth's surface are only a tenth that fast.

Triton, Neptune's largest moon, is one of the most fascinating (and coldest) satellites in the solar system. Imaging from *Voyager 2* revealed "ice volcanoes" spewing frozen nitrogen and dust particles several kilometers into the atmosphere!

In 1990, *Voyager 2* left Neptune and headed south onto a course that has taken it, like *Voyager 1*, to the edge of our solar system and beyond, almost past the reaches of the Sun's winds. This is where interstellar space begins. The spacecraft—fueled by the radioactive decay of plutonium—is expected to continue operating until about 2030.

▶ PHOTO: NASA Jet Propulsion Laboratory

▶ **A PARTIAL VIEW OF TRITON,
NEPTUNE'S LARGEST MOON.**

PHOTO: NASA Jet Propulsion Laboratory

Although its cameras have been shut down to save fuel, *Voyager 2* is still sending fascinating data. Recently it discovered a powerful magnetic field just beyond the solar system. NASA scientists say that hundreds of thousands of years from now, that magnetic field could actually squeeze the area around the solar system, and bring interstellar space closer to Earth. ■

DISCUSSION QUESTIONS

1. Explain how a planet's rings might have formed after the planet was formed. What hypothesis would you make as to where they came from?

2. Since Uranus has a magnetic field, what characteristic(s) might it share with Earth? (*Hint:* Think about atmospheric events.)

PLANETARY FACTS — URANUS

URANUS: QUICK FACTS*

- **Diameter**
  50,724 km
- **Average distance from the Sun**
  2,870,658,186 km
- **Mass**
  8681×10^{22} kg
- **Surface gravity (Earth = 1)**
  0.90
- **Average temperature**
  -195°C
- **Length of sidereal day**
  17.24 hours (retrograde)
- **Length of year**
  84.02 Earth years
- **Number of observed moons**
  27

COMPOSITION

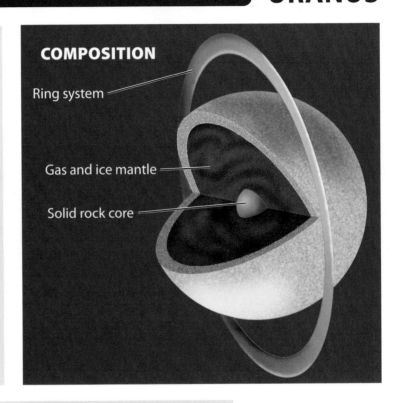

Ring system

Gas and ice mantle

Solid rock core

RELATIVE SIZE

DID YOU KNOW?

- The poles of Uranus are in the same position as the equators on other planets. That's because Uranus rotates on its side.
- It takes nearly 2½ hours for light from the Sun to reach Uranus. (It only takes about eight minutes for the Sun's light to reach Earth!)

ATMOSPHERE

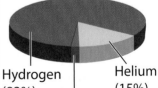

Hydrogen (83%)

Helium (15%)

Methane and traces of other compounds (2%)

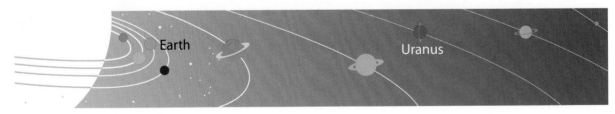

Earth

Uranus

* Source: Data from NASA as of 2011

PLANETARY **FACTS**

NEPTUNE

NEPTUNE: QUICK FACTS*

- **Diameter**
  49,244 km
- **Average distance from the Sun**
  4,498,396,441 km
- **Mass**
  $10,241 \times 10^{22}$ kg
- **Surface gravity (Earth = 1)**
  1.14
- **Average temperature**
  -200°C
- **Length of sidereal day**
  16.11 hours
- **Length of year**
  164.79 Earth years
- **Number of observed moons**
  13

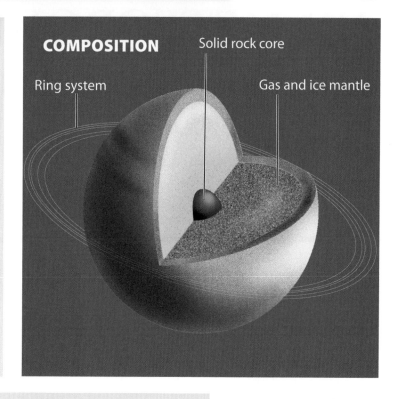

COMPOSITION

Ring system

Solid rock core

Gas and ice mantle

RELATIVE SIZE

DID YOU KNOW?

- Neptune was named after the Roman god of the sea, probably because of its blue color.
- Neptune gives off more heat than it receives from the Sun. This means it probably has its own heat source.

ATMOSPHERE

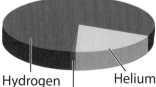

Hydrogen (80%)

Helium (19%)

Methane and traces of other compounds (0.5%)

Earth

Neptune

* Source: Data from NASA as of 2011

ASTEROIDS, COMETS, AND METEOROIDS

COMET HALE-BOPP. THE WHITE DUST TAIL IS COMPOSED OF LARGE PARTICLES OF DUST AND ICE. THE GAS TAIL IS BLUE.

PHOTO: NASA Jet Propulsion Laboratory—Caltech

INTRODUCTION

Planets are not the only bodies orbiting the Sun. On the night of July 23, 1995, astronomers Alan Hale and Tom Bopp discovered a comet while observing from different locations on Earth. Hale and Bopp were the first people in more than 3,000 years to view what is now officially known as Comet Hale-Bopp (Comet H-B for short). This may have been the biggest comet ever visible from Earth. The comet put on a spectacular show as it passed near the Earth on its journey around the Sun.

What other bodies orbit the Sun? In this lesson you will investigate and learn about some other objects that orbit the Sun and are part of our solar system. You will learn about asteroids and comets within the solar system and the possible effects of asteroid and comet impact on the planets. You will also explore what lies beyond Neptune.

What do you already know and want to know about asteroids, comets, meteoroids, and the objects that are in the outer solar system? How are they alike or different? What are some of the possible effects of their impact on Earth and other planets? In this lesson, you will explore these and other questions.

OBJECTIVES FOR THIS LESSON

Analyze the position of the asteroid belt using mathematical patterns.

Explore asteroids, comets, and meteoroids; make comparisons among them.

Read to learn more about objects in the outer solar system.

Summarize and organize information about Pluto.

▶ **MATERIALS FOR LESSON 8**

For you

1   working copy of Student Sheet 2.3c: Solar System Chart

1   completed copy of Student Sheet 8: Bode's Law

GETTING STARTED

1 With your class, review your responses on Student Sheet 8: Bode's Law, which you completed for homework.

2 Read "Asteroids, Comets, and Meteoroids" on pages 161-163. Carefully examine the photos in the reading selection. How are asteroids similar to or different from comets and meteoroids? Record a summary of your ideas in your science notebook. Then discuss them as a class or within your group, as instructed by your teacher. 📝

EXAMINING ASTEROIDS

PROCEDURE

1 In the program *Explore the Planets*, review the Asteroids segment (slides 313-331) in the "Tour the Planets" section. Discuss the concepts with your class.

2 Based on what you have learned in the program, answer the following questions in your science notebook, then be prepared to discuss your ideas with the class: 📝

A. When do scientists think asteroids may have formed?

B. How are asteroids like planets, and how are they different?

C. Why do you think the belt of asteroids exists between Jupiter and Mars?

D. How are the orbits of asteroids similar to, or different from, planetary orbits?

E. How are meteoroids related to asteroids?

▶ THE GOAL OF THE LUNAR RECONNAISSANCE ORBITER (LRO) IS TO MAP THE SURFACE OF THE MOON TO BETTER UNDERSTAND THE HUNDREDS OF THOUSANDS OF CRATERS THAT HAVE RESULTED FROM ASTEROID IMPACTS SINCE IT WAS FORMED.

PHOTO: Chris Meaney, NASA's Conceptual Image Lab

INQUIRY 8.2

EXPLORING COMETS

PROCEDURE

1 In the program *Explore the Planets*, review the Comets segment (slides 341–357) in the "Tour the Planets" section. Discuss the concepts with your class.

2 Based on what you have learned about comets in the program, answer the following questions in your science notebook, and be prepared to discuss your ideas with the class: 📝

A. Where do comets come from in our solar system?

B. What is a comet made of?

C. What is in the tail of a comet?

D. In what direction does the tail of a comet point? Why?

E. Why does a comet glow?

F. How are comet orbits and asteroid orbits alike? How are they different?

G. How are meteoroids and comets related?

INQUIRY 8.3

STUDYING ASTEROID AND COMET IMPACTS

PROCEDURE

1 Look at the photo of the Barringer (Meteor) Crater on page 80. How do you think the crater was formed?

2 Look at the pictures of the asteroids Gaspra and Ida on page 161 and the photos of the Moon and Mercury on pages 80–81. What happened to their surfaces to make them the way they are?

3 Discuss the following questions with your group, then be prepared to share your responses with the class:

A. Are surface impacts a common event or a rare event in our solar system?

B. What evidence do you have to support your response?

4 Watch the video *Dinosaur Extinction*. Discuss the video with your class.

Inquiry 8.3 continued

5 Read "A Fiery Necklace" on pages 164-166. How did Dr. Eugene Shoemaker contribute to the understanding of asteroid and comet impacts? Record your ideas in your science notebook.

6 Based upon what you have learned about impacts on planets, record your answers to the following questions, and be prepared to discuss your ideas with the class:

A. What bodies can collide with other bodies in the solar system?

B. How has Earth's history been influenced by occasional natural catastrophes, such as asteroid impacts?

C. An asteroid impact is considered a natural hazard on Earth, but it is not considered a natural hazard on any other planet or moon. Given this information, how would you define "natural hazard"?

REFLECTING
ON WHAT
YOU'VE DONE

1 Return to your list of ideas about asteroids, comets, and meteoroids. What new things do you want to add to your list? Make your changes and additions now.

2 With your class, return to the Question H (What do you think these objects are? Where would you find them in our solar system?) and Question I (What do you see in the night sky besides stars? Write down what you know about this object.) folders for Lesson 1. Is there anything that you would now change or add? Discuss your ideas with the class.

3 Read "Beyond Neptune" on pages 169–173. Add information from this reading selection to your working copy of Student Sheet 2.3c: Solar System Chart. Discuss the questions that accompany the reading selection with the class.

Asteroids, Comets, and Meteoroids

Among the planets in the solar system are countless asteroids, comets, and meteoroids. Let's take a look at how these solar system objects are different from planets.

ASTEROIDS

Asteroids are metallic, rocky objects in space. They have no atmospheres and move in independent orbits around the Sun. Tens of thousands of asteroids are found in an area called the asteroid belt—a vast, doughnut-shaped ring located between the orbits of Mars and Jupiter (see the illustration below). Gaspra and Ida, pictured at right, are two asteroids found in the main belt. Some scientists theorize that the asteroids in the asteroid belt may be the remnants of an unsuccessful planet. The theory suggests that a planet nearly formed between Mars and Jupiter, but Jupiter's gravity was so powerful that the material could not hold together.

▶ **IMAGE OF ASTEROID 951, GASPRA, TAKEN BY THE *GALILEO* SPACECRAFT**

PHOTO: NASA Jet Propulsion Laboratory/ U.S. Geological Survey

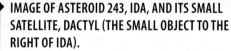

▶ **IMAGE OF ASTEROID 243, IDA, AND ITS SMALL SATELLITE, DACTYL (THE SMALL OBJECT TO THE RIGHT OF IDA).**

PHOTO: NASA Jet Propulsion Laboratory

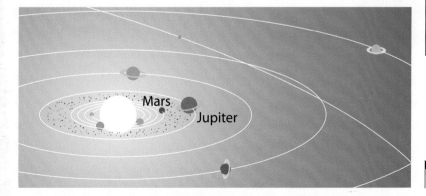

▶ **THE ASTEROID BELT IS LOCATED BETWEEN JUPITER AND MARS.**

READING SELECTION

EXTENDING YOUR KNOWLEDGE

Because asteroids are too small to be classified as planets, they are often called "minor planets." Most asteroids are the size of a small building, but some are much larger. Ceres, found in the asteroid belt, was the first asteroid scientists observed. Discovered in 1801, it is about 1000 kilometers (621 miles) across and is one of the largest known asteroids. Another large asteroid, discovered in 2001, is about a third the size of Earth's moon (or 1150 kilometers). Yet, if we could combine all the asteroids in the asteroid belt now, they would be still be smaller than half the size of the Moon.

COMETS

A comet is a mass of frozen gas, cosmic dust, and ice crystals. Comets are often described as "dirty icebergs." They circle the Sun in long, narrow orbits, mainly located in the cold outer reaches of our solar system. They orbit the Sun in the Kuiper Belt, which begins just past Neptune. A trillion more comets may exist even farther out in a cold area called the Oort Cloud.

Some comets leave their orbits in the Kuiper Belt or the Oort Cloud and journey toward the Sun. When a comet flies near the Sun, its ice begins to "boil" away. As it vaporizes, a tail of glowing gases and dust forms behind it, always pointing away from the Sun. If Earth happens to pass through comet dust, burning particles can be seen streaking through the sky in a spectacular display called a "meteor shower."

METEOROIDS, METEORS, AND METEORITES

Meteoroids are pieces of rock and metal dislodged from comets, planets, asteroids, or moons. Most meteoroids are made up of dust-sized particles. When a meteoroid enters a planet's atmosphere, it burns up due to friction. As it burns, a meteoroid creates a bright streak of light in the sky that we call a "meteor." Sometimes large meteoroids do not burn up completely—one may make it all the way through a planet's atmosphere and land on its surface, after which it is called a "meteorite." Craters on moons and asteroids are evidence of extensive meteoroid bombardment of their surfaces over the course of the solar system's history. ■

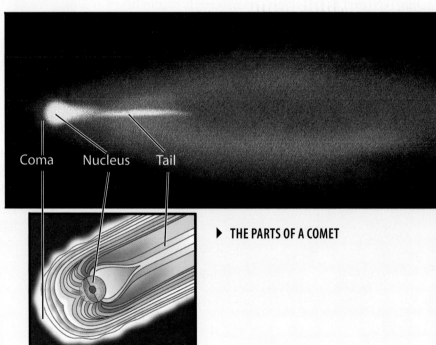

Coma   Nucleus   Tail

▶ THE PARTS OF A COMET

▶ **METEORS STREAK THROUGH THE NIGHT SKY.**

PHOTO: S. Molau and P. Jenniskens, NASA Ames Research Center

▶ **A METEORITE STONE DISCOVERED ON EARTH WEIGHING 452.6 GRAMS. A 1-CM SQUARE CUBE IS SHOWN FOR SCALE.**

PHOTO: NASA Johnson Space Center/Stanford University

DISCUSSION QUESTIONS

1. How would you explain the differences between asteroids, comets, and meteoroids to a young student?

2. Do you think it is accidental that there are so many asteroids in the belt between the orbits of Mars and Jupiter? Why or why not?

READING SELECTION
EXTENDING YOUR KNOWLEDGE

A FIERY NECKLACE

▶ A NASA HUBBLE SPACE TELESCOPE IMAGE OF COMET SHOEMAKER-LEVY 9, TAKEN ON MAY 17, 1994. WHEN THE COMET WAS OBSERVED, ITS TRAIN OF ICY FRAGMENTS STRETCHED ACROSS 1.1 MILLION KILOMETERS OF SPACE, OR THREE TIMES THE DISTANCE BETWEEN EARTH AND THE MOON.

PHOTO: Dr. Hal Weaver and T. Ed Smith (STScI), and NASA

It's not often that you get to see a comet strike a planet. Until July 1994, only one comet strike had ever been observed. That was in 1178, when five English monks reported seeing "a flaming torch" on the Moon "spewing out fire, hot coals, and sparks." Modern astronomers have confirmed that those monks had seen a comet or a small asteroid hit the Moon, forming the crater that is now named Giordano Bruno.

Those 12th century monks weren't equipped with cameras. But in 1994, astronomers all over the world got a chance to see and photograph a similar event when Comet Shoemaker-Levy 9 collided with Jupiter.

BIG NEWS

In 1993, astronomers Eugene and Carolyn Shoemaker, a husband-and-wife team, and David Levy discovered the comet Shoemaker-Levy 9. The three astronomers were working at the Mt. Palomar Observatory in California.

Professional and amateur astronomers look for and discover new comets all the time, but this one made headlines. Shoemaker-Levy 9 was a comet that had been ripped to pieces. Instead of being a single ball, the comet was made up of 21 fragments, each one trailing a large cloud of ice and dust. It looked like a fiery necklace blazing across the night sky. The astronomers calculated that about nine months before they spotted it, Shoemaker-Levy 9 had passed within about 21,000 kilometers (13,049 miles) of Jupiter. Jupiter's gravitational force (which is 2.53 times that of Earth's) had pulled the comet apart.

Next came even bigger news: the pieces of Shoemaker-Levy 9 were on a collision course with Jupiter. Astronomers predicted that Jupiter's gravitational force was about to grab those fragments, once and for all. That set the stage for one of the most photographed events in astronomical history.

RINGING THE BELL

Between July 7 and July 22, 1994, Shoemaker-Levy 9's fragments hit Jupiter's upper atmosphere one by one. Virtually every large telescope on Earth recorded the collisions. The Hubble Space Telescope recorded the event as it orbited Earth, and so did the *Galileo* spacecraft, which was on its way to Jupiter.

The 21 fragments hit Jupiter at speeds of more than 60 kilometers (37 miles) per second. The impacts created plumes of hot gas that

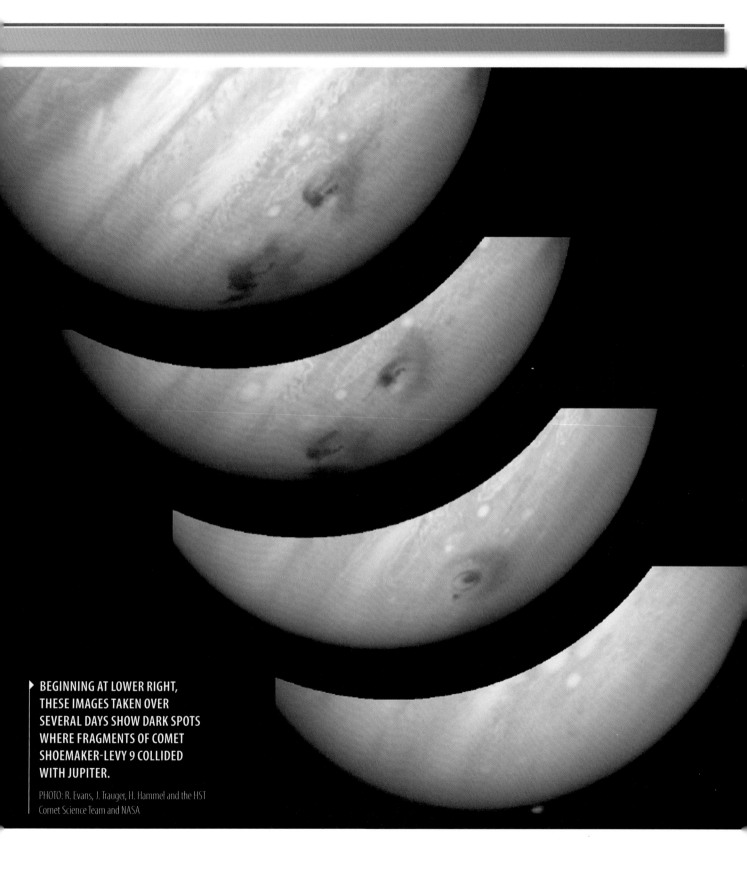

BEGINNING AT LOWER RIGHT, THESE IMAGES TAKEN OVER SEVERAL DAYS SHOW DARK SPOTS WHERE FRAGMENTS OF COMET SHOEMAKER-LEVY 9 COLLIDED WITH JUPITER.

PHOTO: R. Evans, J. Trauger, H. Hammel and the HST Comet Science Team and NASA

TRIBUTE TO EUGENE SHOEMAKER

In July 1997, three years after Shoemaker-Levy 9 collided with Jupiter, Eugene Shoemaker met his own end. He was killed in a car crash in Australia, where he was studying an ancient impact crater.

The following year, Shoemaker's ashes were carried aboard the *Lunar Prospector* spacecraft in a small capsule. In a fitting tribute, the spacecraft was deliberately crashed onto the Moon's surface on July 31, 1999.

"I don't think Gene ever dreamed his ashes would go to the Moon," said his wife, Carolyn. "He would be thrilled."

▶ EUGENE SHOEMAKER AND HIS WIFE, CAROLYN

PHOTO: U.S. Geological Survey

rose thousands of kilometers high. They left marks on the planet's surface that lasted for nearly a year. The pieces plowed into the planet's atmosphere with enormous energy. One astronomer described the impacts as "ringing Jupiter like a bell." ∎

DISCUSSION QUESTIONS

1. Some people—including astronomers—worry that an asteroid or large meteoroid might hit Earth, causing massive devastation. How could you determine whether this is a reasonable worry?

2. Amateur astronomers around the world watched the collision of Shoemaker-Levy 9 with Jupiter, and professional astronomers welcomed their observations. Why does this scientific discipline tend to be open to amateurs?

The Space Name Game

DR. MICHAEL F. A'HEARN

PHOTO: Courtesy of Elizabeth Warner and Dr. Michael F. A'Hearn

Dr. Michael F. A'Hearn knows his way around space. Dr. A'Hearn, a professor of astronomy at the University of Maryland, is also an officeholder in the International Astronomical Union (IAU), the organization responsible for naming celestial bodies. Scientists, space agencies, and authorities around the world recognize and use IAU's names. Here, Dr. A'Hearn answers some common questions about the space "name game," or how astronomical bodies get their names.

Q: How are asteroids named?

A: First, an asteroid is given a set of numbers and letters that tells when it was first discovered. Once the asteroid's orbit is well known, a permanent number is assigned—in numerical order. After that, a name is assigned. The discoverer of the asteroid can suggest a name, but the IAU has final approval.

For example, an asteroid discovered by P. Wild on March 5, 1973, was given the designation "1973 EB." This means that the asteroid was identified in 1973 in the first half of March (E) and was the second (B) asteroid discovered in the first half of that month. Once we understood the orbit of 1973 EB, we gave it the permanent number of 2001 because that's how many asteroids had been discovered by then. This asteroid was named Einstein in memory of Albert Einstein, the greatest scientist of the 20th century.

Q: How often are asteroids discovered?

A: New asteroids are discovered nearly every day! However, people tend not to search during the full moon because the background light interferes too much. Most discoveries are made around the new moon, when our cameras can "see."

READING SELECTION

EXTENDING YOUR KNOWLEDGE

Q: How often are comets discovered? Are they named the same way as asteroids?

A: Many new comets are discovered every month. They are generally named for their discoverers. Comet Halley was named after astronomer Edmund Halley, who was the first to predict the return of this particular comet. But like asteroids, comets are given codes that reflect their discovery date.

Q: What about stars?

A: A few bright stars easily seen from Earth have ancient, traditional Arabic names, such as Sirius. We have identified hundreds of millions of stars. To study them, we need to be able to find them, so they are simply known by catalog numbers. By looking up its number in a huge catalog, we can find a star's precise coordinates, or position in the sky.

Q: Is it true that people can pay to have a star or a planet named after them?

A: Some companies claim to offer such services for a fee. However, those names are completely invalid. As an international scientific organization, the IAU has nothing to do with the commercial practice of "selling" fictitious names of stars, planets, moons, or any other space "real estate."

If you're interested in stars and space, go to your nearest planetarium or local observatory. Have someone show you real stars through a telescope. You also may want to join a local astronomy club. Someday you may discover a new asteroid or comet that could be named after you! ■

DISCUSSION QUESTIONS

1. Dr. A'Hearn says that asteroids are named by their date of discovery. What are other qualities for which an asteroid could be named? What would be an advantage of another naming system?

2. As you've read, the asteroid belt is made up of many asteroids. What advantages might there be to finding and naming all of them?

beyond neptune

after leaving Neptune, *Voyager 2* was not scheduled to fly by any other solar system objects. But there are some interesting objects beyond Neptune that astronomers can observe and hope to send space probes to some day. What lies in the outer regions of our solar system beyond Neptune?

In 1930, astronomer Clyde Tombaugh discovered a new object orbiting the Sun even further away than Neptune. This object was given the name Pluto, after the Greek god of the underworld, and Pluto was considered to be the ninth planet in our solar system. However, as astronomers studied Pluto, they found some characteristics that made it distinct from the other eight planets. For example, the orbit of Pluto is very elliptical, more elliptical than that of any of the other planets. In fact, Pluto's orbit is so elliptical that it sometimes is closer to the Sun than Neptune. They also found that Pluto's orbit is inclined 17 degrees from the plane of the solar system where the other planets orbit.

In 1977, astronomers discovered a moon (Charon) orbiting Pluto. This allowed them to calculate Pluto's mass. It turns out that Pluto is not very massive at all. It is 1/400 the mass of Earth, making it a tiny planet indeed. Some astronomers began to question whether Pluto was a planet or a different kind of object.

Astronomers continued to search the outer solar system using ground-based telescopes and the Hubble Space Telescope, and eventually they found other objects orbiting the Sun far beyond Neptune. In 2005, they identified an object they named Eris. They found that Eris is larger and more massive than Pluto, throwing further into question the idea that Pluto was really big enough to qualify as a planet. Eris is also three

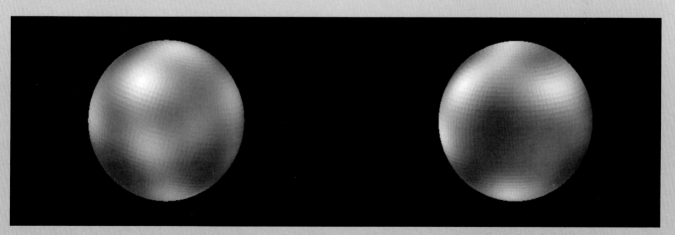

▶ THESE IMAGES OF PLUTO WERE TAKEN WITH THE HUBBLE SPACE TELESCOPE DURING ONE SOLAR DAY ON PLUTO. NOTICE THE LIGHT AND DARK AREAS, WHICH CORRESPOND TO DIFFERENT SURFACE FEATURES.

PHOTO: Alan Stern (Southwest Research Institute), Marc Buie (Lowell Observatory), NASA and ESA

times farther from the Sun than Pluto. Further searches found two more objects—Makemake and Haumea—orbiting the Sun beyond Neptune.

Astronomers call these objects beyond Neptune "trans-Neptunian bodies." Astronomers decided to reclassify Pluto and changed its classification from "planet" to "dwarf planet." Eris, Makemake, and Haumea were also classified as dwarf planets. Pluto, Makemake, and Haumea orbit the Sun in a region called the Kuiper Belt. The Kuiper Belt lies about 50 AU from the Sun. (An AU, or astronomical unit, is the distance from Earth to the Sun, or about 150 million kilometers [93 million miles].) Just as asteroids are mostly found in the asteroid belt, most dwarf planets are found in the Kuiper Belt. Eris itself is too far away to be in the Kuiper Belt, but it's still considered to be a dwarf planet.

The search for Kuiper Belt objects and dwarf planets continues. Other objects have been found and astronomers are working to identify and classify them. Astronomers think there may be as many as 200 dwarf planets in the Kuiper Belt as well as other objects scattered throughout the outer fringes of the solar system.

Even farther out beyond the Kuiper Belt is the Oort Cloud. The Oort Cloud appears to be a spherical cloud of comets about 50,000 AU from the Sun. This cloud of comets is believed to contain the remnants of the material that made up the early solar system. ■

▶ THIS HUBBLE SPACE TELESCOPE IMAGE SHOWS PLUTO AND ITS NEARBY MOON, CHARON, AS SEPARATE, WHEREAS THEY LOOK LIKE ONE OBJECT THROUGH GROUND-BASED TELESCOPES.

PHOTO: Dr. R. Albrecht, ESA/ESO Space Telescope European Coordinating Facility, NASA

READING SELECTION
EXTENDING YOUR KNOWLEDGE

PLUTO FACTS

PLUTO: QUICK FACTS*

- **Diameter**
  2302 km
- **Average distance from the Sun**
  5,906,440,628 km
- **Mass**
  1.3×10^{22} kg
- **Surface gravity (Earth = 1)**
  0.07
- **Average temperature**
  -225°C
- **Length of sidereal day**
  6.39 Earth days
- **Length of year**
  247.92 Earth years
- **Number of observed moons**
  3

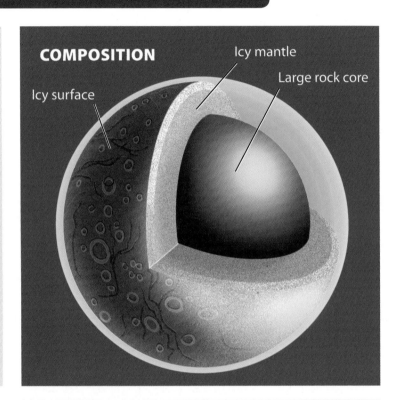

COMPOSITION

Icy surface

Icy mantle

Large rock core

RELATIVE SIZE

DID YOU KNOW?

- Pluto's orbit is more elliptical than the orbit of the planets. Its distance from the Sun varies from less than 4.5 billion kilometers to more than 7 billion kilometers.
- Pluto's orbit around the Sun is tilted 17 degrees, more than any of the planets in the solar system.

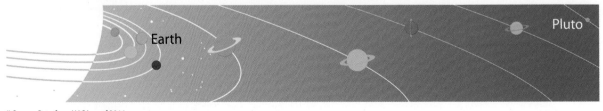

Earth

Pluto

* Source: Data from NASA as of 2011

MISSION TO PLUTO

In 2006, NASA launched a space probe on a mission to Pluto. The space probe is called *New Horizons*. Its task is to photograph and explore Pluto and its moons, Charon, Hydra, and Nix. Because Pluto is 5.9 billion miles from the Sun, it will take *New Horizons* nine years to reach Pluto. In 2015 Pluto will be the first trans-Neptunian object to be visited by a space probe.

▶ AN ARTIST'S IMPRESSION OF THE *NEW HORIZONS* SPACECRAFT APPROACHING AN OBJECT IN THE KUIPER BELT REGION

PHOTO: Johns Hopkins University Applied Physics Laboratory/Southwest Research Institute (JHUAPL/SwRI)

DISCUSSION QUESTIONS

1. Think about what you have learned so far about how we define planets. What about Pluto makes it like a planet and what makes it different?

2. What value do you think there is in continuing to study objects in the far reaches of our solar system?

In 1992, astronomers discovered another interesting object in our solar system. They named this object Chiron. Chiron orbits the Sun but, like Pluto, has a very elliptical orbit. Its orbit brings it closer to the Sun than to Saturn and takes it as far out as Uranus. Chiron has characteristics of an asteroid, such as its chemical composition, and a comet, such as the appearance of a tail. It is therefore hard to classify and has been called a protocomet (extra large comet).

FOSSILS AS EVIDENCE OF ASTEROID IMPACT

INTRODUCTION

An enormous rock falls from the sky—appearing to come from nowhere. Flames of orange engulf the land. All life near the point of impact is destroyed instantly. Particles of melted rock, dust, and debris hover like a dusty blanket and eventually fall back to the planet's surface.

This may sound like a scene from a science fiction movie, but this describes an event that took place again and again and is recorded in Earth's geological history—an asteroid impacting the planet. As you learned in Lesson 8, scientists now think that a large asteroid hit Earth 65 million years ago and caused the extinction of the dinosaurs and many other organisms that inhabited Earth at that time. How do scientists know which organisms died and which survived? They examine the fossils contained in rocks dating from that time. In this lesson, you will explore how fossils help us trace the history of asteroid and comet impacts on Earth.

▶ **AN ASTEROID IMPACT IN WHAT IS NOW PART OF MEXICO MAY HAVE CAUSED THE MASS EXTINCTION OF MANY ORGANISMS, INCLUDING THE DINOSAURS.**

PHOTO: Artist: Don Davis/NASA

OBJECTIVES FOR THIS LESSON

▶ Analyze what the properties of a fossilized rock tell us about how the rock formed.

▶ Brainstorm what you know and want to learn about fossils.

▶ Consider the relationships among fossils, dinosaur extinction, and asteroid impact.

▶ Model fossil excavation, identification, and formation.

▶ **MATERIALS FOR LESSON 9**

For you

1	copy of Student Sheet 9: *Exploring Planetary Systems* Review
1	pair of disposable gloves
1	pair of indirectly vented goggles*

For you and your partner

1	softened fossil-bearing mound
1	plastic wide-mouthed container of warm water, with lid
1	plastic excavating stick
1	paintbrush
1	dropper bottle of water
1	pair of forceps
1	small resealable plastic bag
1	set of fossils (from Inquiry 9.1)
1	plastic wide-mouthed container, with lid (from Inquiry 9.1)
2	shark teeth
1	plastic spoon
1	black china marker
1	plastic wide-mouthed container, with lid (from Inquiry 9.2)
2	shells
1	cup of plaster (premixed by your teacher)
	Wet and dry paper towels
	Newspaper or other table covering

For your group

1	sample of fossiliferous limestone
2	hand lenses
1	"Fossils Through Time" poster
1	cup of gravel, with lid
1	cup of red sand, with lid
1	cup of all-purpose sand, with lid
1	cup of black sand, with lid
1	cup of diluted glue, with lid
1	resealable plastic bag of craft dough

GETTING STARTED

1 Pick up the limestone rock from your group's materials. With your group, examine it with a hand lens. Discuss the properties of the rock.

2 How do you think this rock formed? How did the imprints of the shells get on the rock? Discuss your ideas with the class.

3 Brainstorm with the class what you know and want to learn about fossils.

4 Discuss as a class how the properties of rocks can help scientists learn more about the history of a planet. Consider the following:

A. What did scientists conclude by looking at the layers of the earth in Mexico?

B. Why are rocks and fossils important to scientists who study the history of a planet?

▶ **FOSSIL SHELLS**

PHOTO: Jennifer Aitkens/ creativecommons.org

EXCAVATING FOSSILS

PROCEDURE

1 Discuss what you already know about paleontologists—scientists who study life forms of the past—and how they excavate fossils.

2 Cover your work area with newspaper. Wear gloves. Obtain one fossil-bearing mound for you and your partner. Remove the mound from its container of water. Use your excavating tools (dropper bottle, excavating stick, pair of forceps, and paintbrush) to dig the fossils from your "rock," as shown in Figure 9.1. Work over a paper towel.

3 After you have found the fossils, use your container of water and tools to try to remove as much plaster from them as possible. Dry them. Then clean your work area.

4 Use your hand lens to examine each fossil. Draw a picture of each fossil in your science notebook. Use the fossil identification chart in Table 9.1, your group's fossil poster, fossil books, Figure 9.2, and the class Fossil Collection boxes to identify the name and age of each of your fossils. Write down the name and approximate age of each of your fossils. Create a table to organize your information. ☞

5 Use your black china marker to write your name and your partner's name on a small plastic bag. Place your clean fossils in the bag and seal it.

▶ **USE YOUR EXCAVATING TOOLS TO DIG FOR FOSSILS. BE CAREFUL NOT TO DAMAGE THE FOSSILS.**
FIGURE **9.1**

PHOTO: Courtesy Carol O'Donnell/© NSRC

Inquiry 9.1 continued

TABLE 9.1 FOSSIL IDENTIFICATION CHART

NAME	PICTURE	DESCRIPTION
TOOTH (SHARK)		A common fossil found in marine sediment
MOLLUSK – GASTROPOD		Shell is usually one piece and coiled or spiraled; foot-shaped fleshy material for movement (for example, snail)
MOLLUSK – CEPHALOPOD		A relative of modern squid and octopus; had tentacles; jet-propelled; either coiled or straight shells
BONE (DINOSAUR)		Smooth outer surface; frothy inner surface; bone cells can often be seen
CRINOID STEM		A member of group including starfish and sea urchin; stem made of stacked calcite plates; animal was attached by stem to ocean floor

NAME	PICTURE	DESCRIPTION
CORAL		Small, inactive marine animal, related to sea anemone; has skeleton of calcite material; shape varies (for example, horned, tablet, sponge, branching)
BRACHIOPOD		Type of shellfish; marine animal with clam-like shell; two sides of each shell half are symmetrical
BIVALVE		A member of group including clams and mussels; usually has a pair of identical shells, but two sides of each shell half are not symmetrical
BRYOZOAN		Reef builder; coral-like skeleton formed of calcite; massive or branching colonies; filter feeders
TRILOBITE		One of earliest animals to live in the sea; extinct marine arthropod; three ("tri") body parts; had exoskeleton

INQUIRY 9.2

EXAMINING THE RELATIVE AGES OF FOSSILS

PROCEDURE

1. Discuss with your class how sediments are layered and how organisms become buried in the layers.

2. Examine Figure 9.2 and read its extended caption. Answer the following questions in your science notebook, then discuss them with your class: ☞

   A. Which layer of rock is probably the youngest? Why?

   B. How can the order of rock layers help scientists estimate the age of a fossil?

   C. How can the age of a fossil help scientists estimate the age of a rock layer?

   D. Why is it important for scientists to know the age of a rock or fossil?

3. Put a thin layer (around 1 cm thick) of gravel in the bottom of your wide-mouthed container. (In nature, the layers are called "strata.")

4. Bury the oldest fossil from Inquiry 9.1 in this layer. (Look at your notes to find out which of the fossils is the oldest.) If you don't know the ages, just select any one of your fossils.

5. Now pour a thin layer of sand (any color) into your container until the gravel is covered. Place the next-oldest fossil on top of this sand layer. Harden the sand into rock by covering it with a small amount of diluted glue. (In nature, chemicals seep into sediment and cement the sediment into rock.) Draw a picture of each buried fossil as you work.

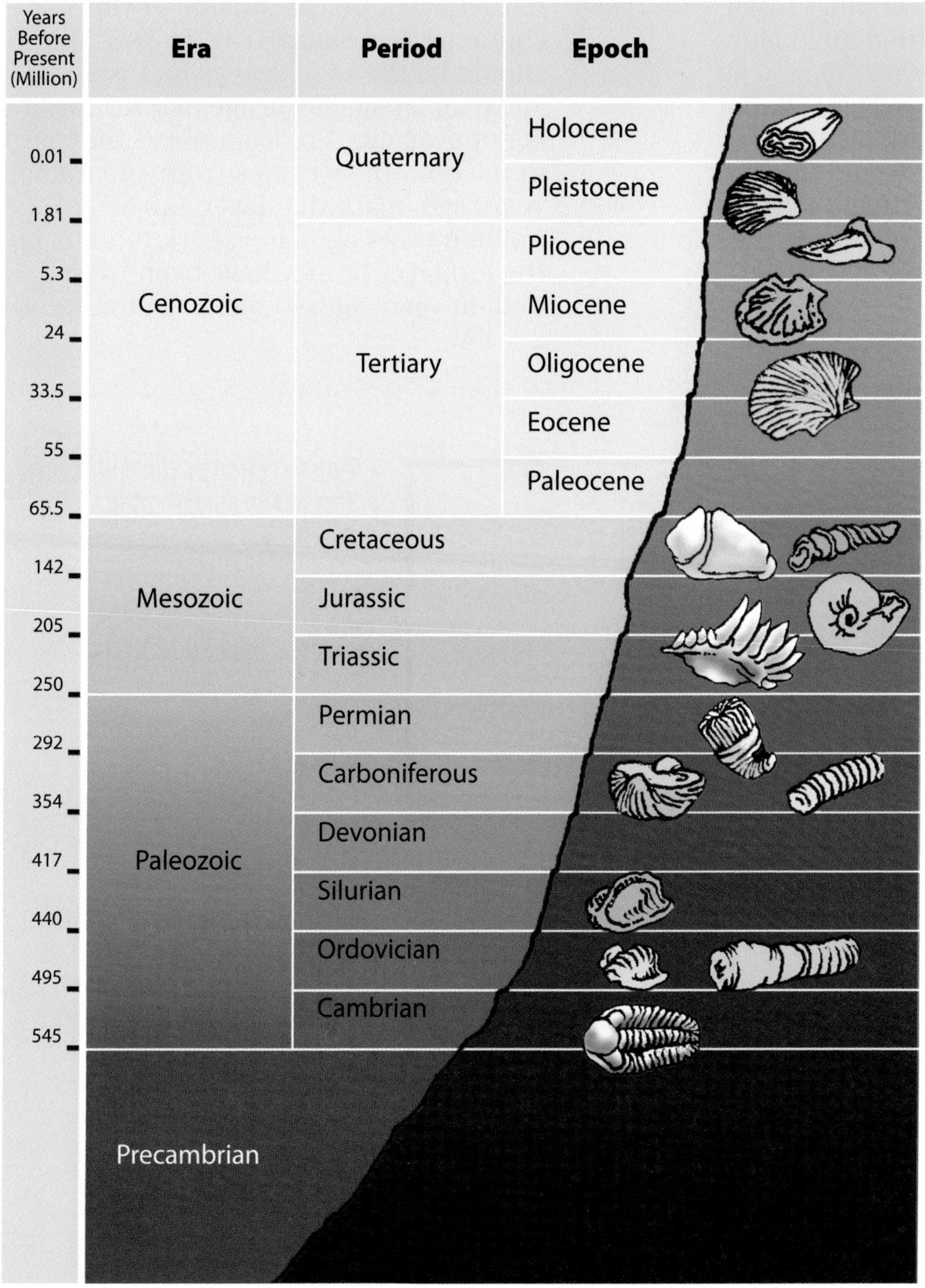

Years Before Present (Million)	Era	Period	Epoch
0.01	Cenozoic	Quaternary	Holocene
1.81			Pleistocene
5.3		Tertiary	Pliocene
24			Miocene
33.5			Oligocene
55			Eocene
65.5			Paleocene
142	Mesozoic	Cretaceous	
205		Jurassic	
250		Triassic	
292	Paleozoic	Permian	
354		Carboniferous	
417		Devonian	
440		Silurian	
495		Ordovician	
545		Cambrian	
	Precambrian		

▶ MUCH OF WHAT WE KNOW ABOUT THE RELATIVE AGES OF ROCKS COMES FROM THE STUDY OF FOSSILS. SINCE CERTAIN TYPES OF FOSSILS ARE ALWAYS YOUNGER THAN OTHERS, WE CAN SAY THAT ONE ROCK IS YOUNGER THAN ANOTHER BASED ON THE FOSSILS IT CONTAINS. IF SEDIMENTS ON EARTH HAD REMAINED UNDISTURBED OVER TIME, THE OLDEST ROCK WOULD ALWAYS BE ON THE BOTTOM AND THE YOUNGEST ROCK WOULD ALWAYS BE AT THE TOP.

FIGURE **9.2**

Inquiry 9.2 continued

6 Repeat Step 5 with another layer of colored sand and another fossil. Harden the layer with diluted glue. Continue this process until all your fossils are buried (see Figure 9.3). Plan to bury the shark teeth in the top layer. (What would the top position of the shark teeth tell you about their age relative to the other fossils? Discuss this with your group.)

7 Use the china marker to write the ages of each layer on the outside of your container (see Figure 9.4). The sediment should be the same age as its fossil. Make up an age if the age of your fossil is not available. For example, if the fossil in the bottom layer was from 440 million years ago, mark that layer 440 MYA (million years ago). If a shark tooth is on the top layer, it may have been from 2 million years ago, so mark the top layer 2 MYA.

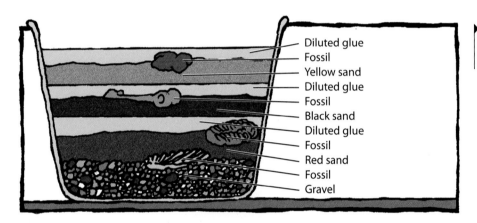

▸ CROSS-SECTION OF CONTAINER WITH STRATA AND BURIED FOSSILS
FIGURE **9.3**

Diluted glue
Fossil
Yellow sand
Diluted glue
Fossil
Black sand
Diluted glue
Fossil
Red sand
Fossil
Gravel

▸ BURY THE FOSSILS IN EACH LAYER. WRITE THE AGE OF EACH FOSSIL ON THE OUTSIDE OF THE CONTAINER. THE OLDEST SEDIMENTARY LAYERS SHOULD BE ON THE BOTTOM.
FIGURE **9.4**

PHOTO: Courtesy of Carol O'Donnell/© NSRC

8 Draw a picture of the final strata. Record the location and approximate age of each fossil.

9 Exchange your container of hardened strata with the other pair of students in your group. Excavate the fossils from the other team's hardened strata, as shown in Figure 9.5. Can you tell how old each fossil is, based on the age of the layer in which you found it?

▸ EXCAVATE ANOTHER GROUP'S STRATA AND DETERMINE THE APPROXIMATE OR RELATIVE AGES OF THEIR FOSSILS.
FIGURE **9.5**

PHOTO: Courtesy of Carol O'Donnell/© NSRC

INQUIRY **9.3**

MODELING MOLDS AND CASTS

PROCEDURE

1 Read "Fossils" on pages 186–189 and discuss with the class the questions that follow. Notice that some fossils are molds (impressions) of an organism, while others are casts. You will model these two fossil processes during this inquiry.

2 Fill the bottom of the plastic wide-mouthed container with half of the craft dough from your group's bag. Flatten the dough.

3 Press one shell, textured-side down, into the dough.

4 Remove the shell from the dough. What does the dough look like? Look back at the reading selection "Fossils." What kind of fossil is an impression of an organism? Discuss this with your group.

5 Repeat Steps 3 and 4 with the other shell if there is room in the dough to do so without disturbing the first imprint.

6 Now use what you know about fossils to create a cast. Fill just the imprint of the shell with plaster as shown in Figure 9.6 and allow the plaster to sit overnight. Label the container with your initials and class period.

7 Once it has hardened, remove the plaster cast from its mold.

▶ MAKING A MOLD AND CAST
OF A FOSSIL
FIGURE **9.6**

1 Share with the class what you've learned about fossil excavation, identification, and formation. Revisit your class list from "Getting Started."

2 Discuss with your class how the molds and casts you made are types of fossils. Why are many fossil molds and casts—like those found in your fossiliferous limestone—made from shells and not from the organism within the shell?

3 Read "The Great Asteroid and the End of the Dinosaurs" on pages 190–193, then answer these questions:

A. On the basis of fossil records, how did an asteroid 65 million years ago affect life on Earth?

B. If the asteroid impact had not occurred, how might life on Earth be different?

C. How might the effects of the asteroid impact have been different on another planet?

4 With your class, return to the Question J (What are fossils? What do fossils reveal about Earth?) folder for Lesson 1. Is there anything you would now change or add? Discuss your ideas with the class.

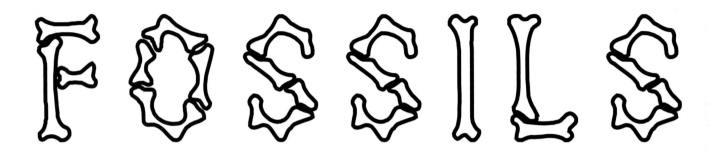

FOSSILS

ossils are the preserved remains of an organism from Earth's past. When a plant or animal dies, it usually decomposes. Bacteria break down its tissue, and over a period of months or years, the tissue disintegrates. But sometimes animal or plant remains are protected from the elements and bacteria that would normally break them down. When this happens, traces of the fleshy parts of the organism may be preserved as a fossil. However, usually only the mineral parts of an organism, such as shells or skeletons, are preserved.

Fossils can be formed in many different ways. One type of fossil is an animal bone, tooth, or shell that has been preserved—often for millions of years. Fossils also can be formed when minerals seep into the pores of a slowly decaying shell or bone and replace the organism's cells with mineral material. Ancient wood preserved in this way is called petrified wood, which is actually stone. An organism preserved in this way does not continue to decay.

▶ IS IT A ROCK OR IS IT WOOD? THIS PETRIFIED WOOD WAS FOUND IN YELLOWSTONE NATIONAL PARK.

PHOTO: NPS photo by J. Schmidt

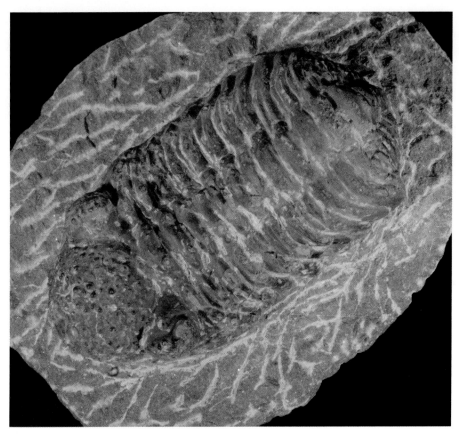

▶ **CAN YOU IDENTIFY THE ORGANISM IN THIS FOSSIL CAST?**

PHOTO: Bruce Avera Hunter/life.nbii.gov

Two other kinds of fossils are molds and casts. A mold is formed when acids dissolve a bone, tooth, or shell and leave an imprint, or mold, of the object in sediment. One way to think about a mold is to imagine pushing a seashell into wet sand. When you pull the shell away, the shape of the shell has been pressed into the sand. If the sand hardened over time, the impression of the shell would be preserved as a mold. A tiny sea creature's skeleton found in a piece of sandstone is an example of a mold. A cast is formed when sand or other minerals fill a cavity-shaped mold over time, and then harden to form a replica of the original organism.

READING SELECTION

EXTENDING YOUR KNOWLEDGE

▶ THIS FLY WAS TRAPPED WHEN IT LANDED ON STICKY RESIN. IN TIME, THE RESIN TURNED TO AMBER AND THE FLY WAS PRESERVED. THIS PIECE OF FOSSILIZED TREE RESIN WAS FOUND IN THE DOMINICAN REPUBLIC AND DATES BACK 24–34 MILLION YEARS.

PHOTO: Courtesy of Carolina Biological Supply Company

Sometimes an entire animal is preserved as a fossil. For example, 30 million years ago a fly may have gotten caught in tree resin. Eventually, that resin hardened into a clear, rocklike substance called amber with the fly entombed inside it. Sediment covered the amber and buried it deep inside the earth. Eventually, erosion may have brought the amber with the perfectly preserved fly to the surface to the lucky person who found it. The oldest amber fossil ever found is more than 300 million years old.

Being trapped in tar also can fossilize animals. In the La Brea Tar Pits in what is today southern California, tar bubbled up to the surface and formed pools in which many animals became trapped. In rare cases, freezing can also preserve an entire organism. Woolly mammoths, woolly rhinoceroses, and musk oxen are animals that were once frozen and have been recovered. Another method for preserving fossils is

▶ THIS FOSSIL RECORDS THE TRACKS OF A THREE-TOED ARCHOSAURUS. IT WAS FOUND NEAR SUMMIT, NEW JERSEY. SEDIMENT WASHED OVER THE TRACKS, AND IN TIME HARDENED INTO STONE, PRESERVING THE TRACKS.

PHOTO: Courtesy of Carolina Biological Supply Company

mummification. Mummification involves the air-drying of soft tissues, such as muscles and tendons, before the organism becomes buried.

Fossils also can include tracks, burrows, borings, nests, or any other preserved indication of the activities of an organism. Even an organism's feces (solid waste) can become fossilized. Fossilized waste can provide important information about diet and the size of the animal that produced it.

The fossilized remains of microorganisms and organisms with skeletons or shells are abundant. Even the fossilized bones and teeth of dinosaurs and sharks are more common than you may realize. The next time you find yourself on a beach, don't just look for shells—look for the fossil of a shell in a piece of sandy rock. You might be surprised at what you find! ■

▶ THIS MAMMOTH WAS TRAPPED IN THE LA BREA TAR PITS OF SOUTHERN CALIFORNIA. SABER-TOOTHED TIGERS ARE IN THE FOREGROUND.

DISCUSSION QUESTIONS

1. What are some of the possibilities for what could happen to an organism after it dies?

2. What sorts of things can fossils tell us about Earth's history as a planet?

READING SELECTION
EXTENDING YOUR KNOWLEDGE

The Great Asteroid and the End of the Dinosaurs

▶ ARTIST'S CONCEPT OF THE ASTEROID IMPACT THAT MAY HAVE BEEN RESPONSIBLE FOR
THE MASS EXTINCTION OF MANY PLANTS AND ANIMALS, INCLUDING THE DINOSAURS.

PHOTO: Artist: Don Davis/NASA

Imagine an asteroid larger than Mt. Everest hurtling toward Earth at a speed of 54,000 kilometers (33,554 miles) per hour. Could anything survive its impact? This scene may sound like something taken from a science fiction tale, but this event really did take place, and many scientists believe it changed life on our planet forever.

IMPACT!

More than 65 million years ago, dinosaurs ruled the earth. Tyrannosaurus rex roamed the plains while winged reptiles soared in the skies above. Strange and exotic vegetation covered the land, while bizarre creatures swam the seas. There was little warning before disaster struck—perhaps just a brief flash across the sky.

The asteroid slammed into Earth with the force of 100 million hydrogen bombs. It struck in the area now known as Chicxulub (cheek-shoe-lube), on the Yucatan Peninsula of what is now Mexico. Trillions of tons of rock vaporized or exploded into the atmosphere. Giant waves known as tsunamis, possibly several kilometers high, destroyed everything in their paths. Trees and plants burst into flames for thousands of miles in every direction. Violent earthquakes shook the planet, while searing winds whipped the landscape. Poisonous gases filled the air. Countless types of creatures and vegetation may have become extinct within hours. Many more would perish within the following days or months, probably because of a lack of habitat or food.

The debris that exploded into the air when the asteroid struck soon spread throughout the atmosphere, blanketing the entire planet in a thick cloud of dust that blocked most of the Sun's rays. Earth was plunged into a cold and terrible darkness. Organisms that could not adapt to the plunging temperature and lack of light began to die.

Plants draw their energy from the Sun and so were probably the first to be affected by the climate changes. Once the plants died, the herbivores (plant-eating creatures) began to starve. And with most of their prey dying off, the carnivores (meat eaters) were forced to attack one another before finally dying themselves. Based on fossil records of the time, scientists believe that the only land animals that

▶ THE ASTEROID HIT WHAT IS NOW THE
YUCATAN PENINSULA IN MEXICO.

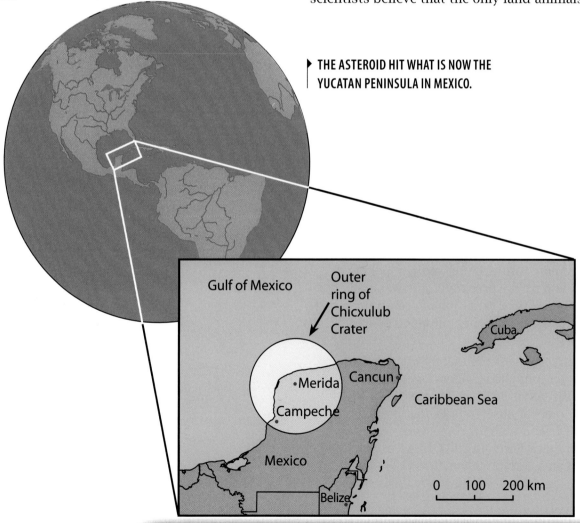

READING SELECTION

EXTENDING YOUR KNOWLEDGE

survived the effects of the asteroid impact were small (weighing 22 kilograms [49 pounds] or less). They probably lived either on insects or on the remains of the larger, dead creatures. Many of them were burrowing animals that could take shelter underground. Larger aquatic animals, such as crocodiles and sharks, did survive the extinction event, perhaps because of a continued food supply of dead materials washed into oceans and streams.

Scientists estimate that when the dust cloud surrounding Earth settled several months later, as few as one third of the organisms living at that time survived to see the Sun again. Most dinosaurs perished, with one exception being those that were the ancestors of modern birds. Smaller mammals, reptiles, fish, amphibians, insects, and plankton survived to become the dominant creatures on the planet.

OUR FUTURE

The question everyone asks is, "Could a disaster like this occur again?" "Yes," says Walter Alvarez. Alvarez is a professor at the University of California at Berkeley and the geologist who—along with his father, award-winning physicist Luis Alvarez, and scientists Frank Asaro and Helen Michel—unlocked the mystery of the great asteroid and dinosaur extinction. Alvarez states, "It can happen and it will happen again in the remote future. But if we keep looking in the sky and we are prepared for it, hopefully our scientists will do a better job of preventing a real natural disaster than the Hollywood people did in the movies."

Looking in the sky is exactly what we're doing. NASA's Jet Propulsion Laboratory runs the Near-Earth Asteroid Tracking project, or NEAT. Astronomers at observatories that participate in NEAT keep watch for comets and asteroids that approach Earth. In 2002, NEAT astronomers spotted an asteroid 1000 meters (3281 feet)

in diameter that looked to be headed straight for Earth. Careful calculations, though, showed that there was no need for panic: it would come close to Earth, about twice the distance of the Moon, but miss. It was a good thing, too. Had the asteroid collided with Earth, astronomers said, it could have wiped out an area the size of Texas.

What might happen if an asteroid really were on a collision course with Earth? Right now, the answer is, "We're not sure." Scientists are exploring asteroids to find out whether we might be able to nudge them out of their trajectories, making them miss Earth. They're also learning more about asteroids in order to see whether we could blow up large asteroids headed for Earth, and turn them into smaller, less harmful debris.

This, when you think about it, is remarkable work for a species that only came into existence about 200,000 years ago. Even though dinosaurs inhabited Earth for far longer than humans have, we are the first species on the planet that may have the ability to save itself from such a rare and random threat coming from outer space. ■

DISCUSSION QUESTIONS

1. If an asteroid impact had not occurred 65 million years ago, how might life on Earth be different?

2. How do you think scientists can figure out how we might nudge or blow up asteroids that threaten to collide with Earth?

HOW DO WE KNOW WHAT HAPPENED 65 MILLION YEARS AGO?

In 1980, scientists made an educated guess about the location of the impact crater at Chixculub after studying differences in mineral deposits in the soil. In 1990, a NASA-generated image from space confirmed the existence of a crater in that area. Over the course of 65 million years, the crater has been nearly filled with soil and sediment. But a few telltale clues helped scientists unlock the mystery.

First, quartz rock in the area showed signs of great stress that result from a meteorite impact. This quartz is known as "shocked quartz." Second, large amounts of a mineral called iridium, which is found only in meteorite dust or debris from an asteroid impact, was detected at Chixculub.

Meanwhile, paleontologists were studying fossil records from Chicxulub and other places around the world. When they reached a certain layer in the strata, paleontologists could see a dramatic change in the types of fossils in the soil. More importantly, between the layers of strata where the dramatic change occurred, scientists found a thin layer of iridium. The fossils that lay beneath the iridium are thought to be from organisms that lived before the impact. The fossils above the iridium are from organisms that either survived the impact or evolved from them. Working as "time detectives," scientists were able to determine that nearly two-thirds of all organisms on Earth vanished at the time of the asteroid impact.

▶ A LAYER OF IRIDIUM IS EVIDENT AS A DARK BAND. LAYERS BELOW THE DARK BAND WERE LAID DOWN BEFORE THE IMPACT. LAYERS ABOVE THE DARK BAND WERE LAID DOWN AFTER THE IMPACT. THE POCKET KNIFE IS SHOWN FOR SCALE.

PHOTO: Courtesy of Professor David A. Kring

The Age of Planets: Dating Rocks

ALL ORGANISMS ABSORB CARBON FROM THE ENVIRONMENT. THIS CARBON DECAYS, BUT IS REPLACED BY NEW CARBON UNTIL THE ORGANISM DIES. AFTER DEATH, THE CARBON IN THE ORGANISM CONTINUES TO DECAY AT A STEADY RATE, WHICH TELLS US HOW LONG AGO THE ORGANISM DIED.

Scientists believe that our solar system is about 4.5 billion years old. How did they make that estimate? Scientists use a process called radiometric dating to help them determine the age of rocks that are found on Earth and the Moon and to date meteorite samples. By comparing the ages of these materials, scientists are able to draw conclusions about the age of the entire solar system.

RADIOMETRIC DATING

Radiometric dating uses radioactive elements to help determine the age of rocks (and sometimes fossils) by measuring the age of radioactive elements in them. It works like this: Radioactive atoms are naturally unstable, and release subatomic particles—neutrons, electrons, and protons. As they do this, they can become, or "decay into" other elements. Radioactive carbon, for instance, can turn into nitrogen: we call nitrogen the "decay product."

In any given sample of material, we can expect a certain percentage of each element's atoms to be radioactive. By checking to see how much, say, radioactive carbon is in a sample, and

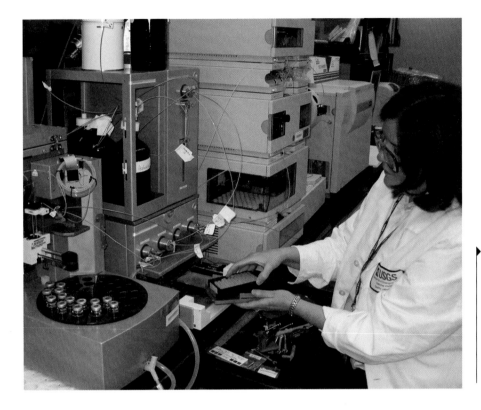

▶ A TECHNICIAN OF THE U.S. GEOLOGICAL SURVEY USES A MASS SPECTROMETER. A MASS SPECTROMETER ALLOWS SCIENTISTS TO PRECISELY ANALYZE ROCKS AND MINERALS RANGING IN AGE FROM 10,000 YEARS TO 0.56 BILLION YEARS.

PHOTO: Jon Raese/U.S. Geological Survey

comparing it with what should be there, we can see how much has already turned into nitrogen. Once we know that "missing" quantity, we can calculate the amount of time it must have taken for that much radioactive carbon to turn into nitrogen. That tells us how old the substance is.

Scientists can measure the amount of radioactive carbon in dead or fossil organisms to calculate when the organism died. This method works for organisms that lived up to 70,000 years ago. Radiometric dating won't work on living organisms, though. It assumes that at a certain point in time, the substance stopped changing: No new material was brought into it, no material left it, its radioactive elements were left to decay in peace, and their decay products were allowed to build up.

Let's take a closer look at how radiometric dating works, using carbon dating as an example.

The process of carbon decay does not happen quickly. In fact, it takes 5730 years for just half the radioactive carbon in a sample to decay into nitrogen. This time is called the "half life" of radioactive carbon. (Actually, there are several types of radioactive carbon, and some of them decay within minutes or hours. The carbon you'll hear about most often in radiometric dating is called carbon-14, and this is the one with the long half life.)

The concept is pretty simple for the first half life, but it gets more complicated after that. Imagine that you have a 100-gram sample of radioactive carbon. If you could check on that carbon after 5730 years, you would find that half the carbon had decayed into nitrogen. You would have 50 grams of radioactive carbon and 50 grams of nitrogen.

▶ OUTCROP IN WESTERN AUSTRALIA. ZIRCON CRYSTALS (SEE BELOW) NEAR THE HAMMER WERE DATED AT 4.4 BILLION YEARS OLD. THESE CRYSTALS ARE PIECES OF THE EARLIEST KNOWN CRUST ON EARTH. (FROM LEFT TO RIGHT: JOHN VALLEY AND AARON CAVOSIE, UNIVERSITY OF WISCONSIN–MADISON, AND SIMON WILDE, CURTIN UNIVERSITY)

PHOTO: Courtesy of John W. Valley, University of Wisconsin–Madison

If you could come back after another 5730 years (or after a total of 11,460 years) you would find that half the carbon that was left the previous time had decayed, so you would have 25 grams of radioactive carbon left. Then, 5730 years later, you'd have 12.5 grams of radioactive carbon left. As time passed, the amount of radioactive carbon would get smaller, but it would take a very long time for all the carbon to decay.

Radiometric dating is extremely useful when combined with the kind of fossil dating you've done in your lab: relative dating. While it's fine to know that one fossil is older than another, that by itself can't tell us exactly how long ago the creature lived. If one earth layer's fossils are 2.2 million years old, though, and a higher layer's fossils are 1.3 million years old, it's a reasonable assumption that all fossils found

▶ MICROSCOPIC VIEW OF A ZIRCON CRYSTAL DETERMINED TO BE 4.4 BILLION YEARS OLD

PHOTO: Courtesy of John W. Valley, University of Wisconsin–Madison

between the layers are between 1.3 and 2.2 million years old. That helps us group fossils around the world into their respective eras and periods, and helps give a more complete picture of the entire Earth during each time.

Scientists who want to study things as old as rocks need to select a radioactive element that has a much longer half-life than that of radioactive carbon, such as radioactive potassium, or potassium-40. Potassium-40 turns into argon as it decays, and its half-life is 1,280,000,000 years (or 1.28 billion years). By comparing the amount of radioactive potassium to the amount of argon in a rock sample, they can determine the age of the rock. Several other elements, or combinations of elements, can also be used to date very old objects.

In western Australia, the ratio of lead to uranium was used to date crystals called zircons; in zircons, this ratio increases with age. Crystals of zircon were dated at more than 4.2 million years, making them the oldest rocks found so far on Earth. Within the zircons, tiny, ancient diamonds were found. While the diamonds were too small to have any commercial value, they are invaluable for understanding what Earth used to be like. Some scientists point to their structure as evidence that cooler temperatures, liquid oceans, and rocky continents set the stage for life very early in Earth's history.

THE ANCIENT SOLAR SYSTEM

In their efforts to determine the age of the solar system, scientists have not limited themselves to studying Earth. They have also studied rocks from the Moon. Many of these samples were collected during NASA's Apollo missions. Scientists have determined that the oldest of these lunar rocks are between 4.4 and 4.5 billion years old, which indicates that the Moon is about as old as Earth.

Scientists also have studied meteorites to determine the age of the solar system. They help us get around a problem in dating very old Earth rocks: Ancient rocks deep in Earth's crust may undergo intense heat and pressure, causing them to change into "metamorphic" rocks. This sometimes resets the radiometric clocks, such that the metamorphic rocks show up only as old as the time of their metamorphosis. The age of their original "parent" rocks is no longer detected. However, we can find ancient meteorites that have fallen to Earth and never undergone metamorphosis. We can trust them as good indicators of the age of the solar system. Using radiometric dating, scientists have now determined the age of more than 70 meteorites, all of which are about 4.5 billion years old.

As scientists continue to study new mineral samples from other planets, it will be possible to test rocks from other places in the solar system. Those samples may allow scientists to give us a more precise idea of the age of the solar system and how it has developed since its birth. ∎

DISCUSSION QUESTIONS

1. How have scientists come up with estimates of the age of our planet and other parts of the solar system?

2. Why is it important for us to know the ages of the planets, including our own?

little things Mean A LOT

Throughout this unit, you read a series called "Missions." The information about Earth's solar system is very detailed—and perhaps confusing at times.

"Why does this all matter?" you might ask. "Is it really important that Earth is 149.6 million kilometers (about 93 million miles) from the Sun? Or that Earth's surface gravity is much less than that of Jupiter? Or that the Moon is 384,400 kilometers (238,855 miles) from Earth?"

These details do matter. They matter because it's the balance of many factors—mass, distance from the Sun, rotation rate, and others—that makes life as we know it possible on Earth. These factors also make Earth, the other planets, and the Sun work together as a whole—as a system.

Two scientists, Stephen Dole and Isaac Asimov, described the importance of little things in *Habitable Planets for Man*, a book published in the 1960s. In it they speculate about life on Earth if some things were changed. Let's examine a few.

WHAT IF EARTH WERE TWICE AS MASSIVE?

Greater mass would mean a greater surface gravity. This would have a significant effect on plant and animal life. Trees would be shorter and have thicker trunks. Animals would have heavier leg bones and muscles. Mountains would not be as high because they would not have the strength to support their weight. Waves in the ocean would be lower, and erosion would be faster.

WHAT IF EARTH WERE CLOSER TO THE SUN?

If Earth's mean distance from the Sun were 10 percent less than it is now, less than 20 percent of Earth's surface would be habitable. The habitable areas would lie in two bands between latitudes 45 degrees North and 64 degrees South. A broad area of intolerable heat would separate these two bands. There would be no polar ice, and the level of the oceans would be higher.

WHAT IF EARTH ROTATED ONCE EVERY 100 HOURS, RATHER THAN EVERY 24 HOURS?

Temperature differences between day and night would be extreme. The Sun would seem to crawl across the sky. Few forms of life would be able to tolerate both the intense heat of the long days and the bitter cold of the long nights.

WHAT IF THE MOON WERE MUCH CLOSER—FOR EXAMPLE ABOUT 152,000 KILOMETERS (94,448 MILES) AWAY FROM THE EARTH, INSTEAD OF 384,400 KILOMETERS (238,855 MILES) AWAY?

If the Moon were much closer to Earth, tidal forces might be strong enough to halt the rotation of Earth with respect to the Moon. A "day" on Earth would last a month, and Earth would be uninhabitable. By contrast, if the Moon were more than 713,600 kilometers (443,410 miles) away, Earth could not hold it in orbit. Organisms dependent on the rhythm of the tides would die out, and nights would always be dark.

As we can see from these examples from *Habitable Planets for Man*, life on Earth is possible only because of a delicate balance that exists in the solar system. Change just one thing and we might be in for trouble. Details do matter! ■

DISCUSSION QUESTIONS

1. What characteristics of Earth allow life to exist on the planet? How might Earth change if any of its conditions were altered slightly?

2. Examine the planet you studied for your planetary travel brochure. What conditions would you have to change for life to exist on that planet?

SOLAR SYSTEM ASSESSMENT

▶ **HOW DOES A PLANET'S DISTANCE FROM THE SUN AFFECT ITS ORBITAL PERIOD?**

PHOTO: Jeff McAdams, Photographer, Courtesy of Carolina Biological Supply Company

INTRODUCTION

You have now completed your exploration of the solar system. In this lesson will assess your understanding of the concepts and skills related to your study of the solar system in Lessons 1–9.

This assessment is divided into three parts. For Part A, you will evaluate an investigation, analyze data, and draw conclusions. You also will be asked to describe how the investigation relates to the solar system. For Part B, you will complete multiple-choice and short-answer questions about the solar system. For Part C, you will revisit the questions in Lesson 1 and respond to them again. This will give you an opportunity to self-assess what you have learned in this unit.

OBJECTIVES FOR THIS LESSON

Review and reinforce concepts and skills from *Exploring Planetary Systems*.

Complete a three-part assessment of the concepts and skills addressed in *Exploring Planetary Systems*.

▶ **MATERIALS FOR LESSON 10**

For you

1   copy of Inquiry Master 10a: Solar System Assessment Experiment Sheet (Part A)

1   copy of Inquiry Master 10b: Solar System Written Assessment (Part B)

1   copy of Student Sheet 10a: Solar System Assessment Experiment Analysis Sheet (Part A)

1   copy of Student Sheet 10b: Solar System Written Assessment Answer Sheet (Part B)

GETTING STARTED

1 Discuss your responses to Student Sheet 9: *Exploring Planetary Systems* Review with your group and then with the class.

2 Ask questions about anything you do not understand on Student Sheet 9.

▶ **WHAT HAVE YOU LEARNED ABOUT THE SOLAR SYSTEM FROM THIS UNIT?**

PHOTO: NASA Jet Propulsion Laboratory

PART A

PROCEDURE

1 Read the first paragraph on Inquiry Master 10a: Solar System Assessment Experiment Sheet (PartA). This describes what you will be doing for Part A of the assessment. Ask questions if you do not understand what is expected for this part of the assessment.

2 On Student Sheet 10a: Solar System Assessment Experiment Analysis Sheet, write a hypothesis for the question you are investigating.

3 Study the experiment described on Inquiry Master 10a, then complete Student Sheet 10a as directed.

4 Turn in completed Student Sheet 10a and Inquiry Master 10a when you have finished.

PART B

PROCEDURE

1 Complete Part B. Do not write on Inquiry Master 10b: Solar System Written Assessment (Part B); other classes throughout the day will use it. Write answers on Student Sheet 10b: Solar System Written Assessment Answer Sheet (Part B).

2 When you have finished, turn in Inquiry Master 10b and Student Sheet 10b to your teacher.

PART C

PROCEDURE

1 Review Parts A and B of the assessment with your teacher.

2 Revisit Lesson 1 and answer the questions again, then compare your responses with your science notebook entries when you began the unit. How have your ideas about the solar system changed since Lesson 1? 🖉

HERITAGE IN SPACE

n 1992, a team of NASA scientists set out on what looked to some like a lunatic mission. They were turning a huge radio-telescopic ear to the cosmos, scanning hundreds of star systems, and hoping to hear…someone. Or something. They were listening for signals from space, even signals we couldn't understand. If only we could recognize them as signals, we'd know that something was out there broadcasting to the universe, and that we were not alone. This project was NASA's contribution to SETI, the Search for Extra-Terrestrial Intelligence. First conceived by a group of enthusiastic scientists in the 1960s, SETI is still running today, still listening to the cosmos. To date, we appear to be all alone in the universe.

SETI raises hard questions. Does it make any sense to listen to space? How likely is it that intelligent life exists anywhere else? Could it have started anywhere but on Earth?

▶ **IS THERE LIFE AMONG THE STARS IN OUR GALAXY?**

PHOTO: NASA, ESA, and The Hubble Heritage Team (STScI/AURA)

The scientific field that deals with these questions is called astrobiology, and it has tantalized astronomers since its origins in the middle of the 20th century. In 1953, Stanley Miller and his teacher, Harold Urey, showed that the chemical soup of the early Earth, when struck by lightning, could have formed

RADIO TELESCOPE IN CALIFORNIA

PHOTO: Colby Gutierrez-Kraybill/creativecommons.org

READING SELECTION

EXTENDING YOUR KNOWLEDGE

amino acids and other molecules essential to Earth life. These are organic compounds, meaning that they are carbon-based. They're also biochemicals: organic compounds such as nucleic acids, amino acids, sugars, fats, and a vast array of larger molecules built from them. Some of these molecules are small, some are large, but they have carbon backbones in common, and—most important—they are either self-reproducing or part of systems that can reproduce themselves: in other words, something like life.

Not long after the Miller-Urey experiments, chemical analysis of a famous 19th-century meteorite showed that it carried organic compounds, too. They weren't biochemicals, but they might have been precursors. Precursors are molecules that can become biochemicals through chemical reactions. A small group of scientists—prominent chemists, biologists, physicists, and astronomers, including the young Carl Sagan—had already been probing the question of whether extraterrestrial life existed. Now they were seized by the idea that

long ago, such biochemical precursors had been brought to Earth by meteorites, and eventually evolved into the molecules of life: DNA, RNA, amino acids, carbohydrates, and others. These scientists began a dialogue about how molecules from space might have seeded Earth with biochemical precursors. If these basic molecules did come from space, how did they get here, and in what form, and how did they evolve into DNA, RNA, amino acids, and other biochemicals? And could this really have happened? Could life on Earth really have gotten its start in space?

Over the next few decades, their work on chemical evolution merged with other scientific work being done on the origins of life on Earth, and grew into the field called astrobiology. To explore its questions, if not answer them definitively, NASA set up its astrobiology division, where scientists research everything from the life and chemistry of early Earth oceans to the possibility of organic molecules floating through space and "seeding" planets with the stuff of life.

▶ CLOUDS OF DUST AND GAS CONDENSED TO FORM THE SUN AND EIGHT PLANETS.

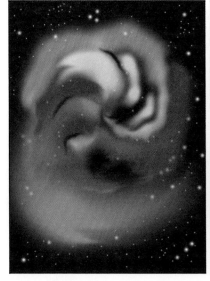 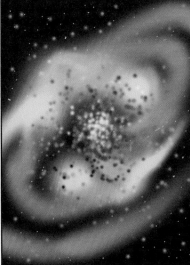 

What have they found? They suggest that within interstellar clouds, before planetary systems form, atoms and molecules swirl, reacting with each other. Though relatively rare in space, organic compounds do form. Some of them are drawn into gathering the clumps of matter that become stars, planets, and moons. Others drift through space, continuing to react and evolve, and being gathered slowly into ices, dust, and other small clumps of space matter.

Astrobiologists now think that interstellar dust, meteorites, and comets may have rained down on Earth very early in Earth's history, dumping organic chemicals into the young Earth seas, and giving the planet a good stock of biochemical precursors. Some of them may even have evolved into biochemicals by the time they reached Earth.

It's not enough, though, just to have a lot of organic chemicals, or even biochemicals, floating around. By itself, that doesn't make life. The molecules have to be able to start making copies of themselves, and join together in ways that lead to organismal life: bacteria, plants, animals, and the like. We don't yet know how this happened on Earth, which means that we don't know what conditions might be necessary for it to happen elsewhere. And just as we are still trying to discover what early hominids looked like, we don't yet know what exactly our early biochemical precursors might have been. If we can figure that out, we can narrow our search for life to those planets with the right conditions, and the right molecules.

Even then, we would still be a long way from finding extraterrestrial creatures who can broadcast meaningful signals across the universe. Many other questions arise: What planetary conditions might be necessary for biochemicals to survive, join together, start reproducing, and develop into organisms? What's necessary for the development of a life form that's robust enough, and intelligent enough, to communicate with others on their own planet? And how do they grow intelligent and capable enough to put out signals—intentionally or not—to the rest of the universe?

In asking these questions we find, suddenly, that we still don't have answers to the very simplest questions. Here on Earth, we have a pretty good idea of what we mean when we say "life." We think of life as requiring liquid water, oxygen, and various other conditions. But what's to say that life must develop this way on every planet? Can biochemicals find a way to reproduce on a dry planet? Can intelligent life develop without lots of atmospheric oxygen? Do we really know what an "intelligent life form" is? For that matter, does our definition of "life" really make sense when we scan the universe? Is there some other kind of "life" that can communicate? Maybe so—but if we can't figure out what it could reasonably be, we'll never look for it.

So there it is: we search the universe for signs of life, and the universe pushes our gaze back on ourselves, and makes us think harder about what life is. ■

DISCUSSION QUESTIONS

1. What are some conditions that may have to exist on a planet for life to form?

2. How does the search for extraterrestrial intelligence call into question our understanding of what life is?

Glossary

amber: A type of fossil that forms when partial or complete insects and small arthropods become embedded in tree resin; fossilizes when the resin is buried and hardens into a clear shell. See also *fossil*.

asteroid: A small, mostly rocky solar system object that orbits independently around the Sun; minor planet. See also *asteroid belt*.

asteroid belt: A large group of asteroids that orbits the Sun between Mars and Jupiter. See also *asteroid*.

astronomer: A scientist who studies the stars, planets, and other objects in space. See also *astronomy*.

astronomical unit: A unit of measure equal to the average distance between Earth and the Sun, about 150 million kilometers (93 million miles); abbreviated AU.

astronomy: The branch of science that studies the stars, planets, and other objects in space. See also *astronomer*.

atmosphere: The mixture of gases that surrounds a planet or moon.

axis (plural: axes): An imaginary line that runs through the middle of an object (for example, from pole to pole) around which that object rotates; a line at the side or bottom of a graph.

basin: An area where rock dips toward a central point or depression, as in a crater. See also *crater*.

cast: Type of fossil that forms when sand, minerals, or other matter fill a cavity-shaped mold over time and then harden, forming a replica of the original organism. See also *fossil*.

celestial: Of or relating to things in the heavens.

coma: The part of a comet that surrounds the nucleus and is made of gas and dust. See also *comet*.

comet: A celestial body composed of a mixture of rocks, frozen gases, cosmic dust, water ice, and organic compounds that orbit the Sun in a variety of elliptical paths that pass close to the Sun and swing to the outer reaches of the solar system. Occasionally, a comet may pass close enough to a larger planetary body whose gravitational field captures the comet, causing a collision with the more massive body.

constellation: An observed pattern of stars.

core: The center of a planet, star, or moon.

crater: A bowl-shaped pit on a planet, moon, or asteroid formed by the impact of an object; also formed by volcanoes. See also *basin*.

Earth-centered: A description of the universe in which it was believed that all the planets, stars, the Moon, and the Sun revolve around Earth.

ellipse: An oval-shaped closed curve; the shape of a planet's orbit.

erosion: The process by which terrestrial planetary materials are broken down and moved from place to place, for example, by wind and water.

flyby: Method astronomers use to observe a planet or moon whereby a spacecraft "flies by" the planet or moon, taking pictures of it and gathering other scientific data as it does.

fossil: The preserved remains or impressions of organisms of Earth's geological past. See also *amber, cast, mold.*

galaxy: A large system of dust, gas, stars, and other celestial bodies that has a particular shape.

gaseous planets: Planets composed of compounds that under normal Earth conditions would be gases; includes Jupiter, Saturn, Uranus, and Neptune.

gravity: A force of attraction between two objects. See also *Law of Universal Gravitation.*

gravity assist: A technique that uses the pull of a planet's gravity to change a spacecraft's speed and direction.

greenhouse effect: The trapping of heat by a planet's atmosphere.

greenhouse gases: The gases in a planet's atmosphere, such as water vapor and carbon dioxide, that absorb energy radiated from the planet and prevent its escape into space.

Hubble Space Telescope: A telescope that orbits 600 km above the surface of Earth.

inertia: The tendency of an object to remain either at rest or in motion unless acted on by an outside force. See also *Law of Inertia.*

lander: A spacecraft that lands on a planet or moon to gather data directly from its surface.

landform: A physical feature of a planet's surface, such as a mountain, plain, or valley.

latitude: An angular distance on a globe that runs parallel (east and west) to the equator; measured in degrees north and south of the equator. See also *longitude.*

Law of Inertia: Law stating that a body in motion tends to travel at a constant speed in a straight line unless an outside force disturbs it. See also *inertia.*

Law of Universal Gravitation: Law stating that any two objects in the universe have gravity and will attract each other, and that attraction depends on how much mass each object has and their distance from each other. See also *gravity.*

longitude: An angular distance on a globe that runs perpendicular (north and south) to the equator; measured in degrees east and west. See also *latitude*.

maria: Dark, flat, low-lying regions on the Moon's surface.

mass: The total amount of matter in an object; not dependent upon gravitational pull. See also *weight*.

meteor: The streak of light that is produced when a meteoroid burns as it enters an atmosphere. See also *meteoroid*; *meteorite*.

meteorite: A meteoroid that strikes a planet, moon, or asteroid. See also *meteor*; *meteoroid*.

meteoroid: A solid object moving in interplanetary space, distinguished from asteroids and planets by its smaller size. See also *asteroid*; *meteor*; *meteorite*.

model: A representation that is used to study objects, ideas, or systems that are too complex, distant, large, or small to study easily firsthand.

mold: A fossil type that is an impression of a shell, bone, tooth, or other body part left in the rock after the organism is covered by soft material. See also *fossil*.

moon: A rocky object that orbits a planet; a natural satellite.

NASA: The National Aeronautics and Space Administration, an organization that oversees the United States' space program, established in 1958.

nebula: A concentration of dust and gas in space.

nuclear fusion: The reaction by which hydrogen gas changes into helium gas and releases energy in the form of heat and light.

nucleus: The main part of a comet, which is made of ice, gas, and dust.

orbit: (noun) The curved path of one object, such as a planet or moon, around a central object, such as a star or planet; (verb) to move in a circular or elliptical path around a central object. See also *revolve*.

orbital period: The time that it takes an object to orbit another object one complete time. See also *period of revolution*.

orbiter: A spacecraft that studies a planet by orbiting it rather than by flying past it.

paleontologist: Scientist who studies life forms of the past. See also *fossil*.

period of revolution: The time it takes an object to orbit another object one complete time. See also *orbital period*.

period of rotation: The time it takes an object to spin on its axis in one complete rotation. See also *rotation*.

petrified wood: Fossil originally of wood in which the wood has been replaced by some mineral. See also *fossil*.

plane: A flat surface; an imaginary surface along which the planets orbit.

planet: A massive, usually spherical space object that orbits a star and shines by reflecting the star's light.

probe: Instrument that makes observations and takes measurements such as atmospheric content, turbulence, temperature, particle size, and radiation either on a planet's surface or in its atmosphere.

radiation: The process by which energy is transferred from one object, such as the Sun, to another object, such as a planet, without the space between them being heated.

rays: Spoke-like patterns of ejected material that radiate from a crater.

revolution: The movement of one object around a central object. See also **revolve**.

revolve: To move in a curved path or orbit around a central object. See also *orbit*; *revolution*.

rotate: To turn or spin around a central point or axis. See also *axis*; *rotation*.

rotation: The movement of one object such as a planet or moon as it turns or spins around a central point or axis. See also **rotate**.

satellite: A natural (for example, the Moon) or artificial (for example, the Hubble Space Telescope) object that orbits another object in space.

scale: The ratio between the measurements on a map or model and the actual proportions of an object.

scale factor: A method for reducing all measurements by the same amount to achieve the measurement of the scale model.

solar system: A star with planets and other objects in orbit around it; our solar system is made up of the Sun, eight planets, asteroids, meteoroids, comets, and other space objects.

space probe: An unmanned spacecraft that collects information in space.

space shuttle: A reusable spacecraft designed to transport astronauts, materials, and satellites to and from Earth's orbit.

star: A sphere of hot glowing gases that releases energy in the form of heat and light. See also *Sun*.

Sun: The star in the center of our solar system around which Earth and seven other planets revolve. See also *planet*.

technology: The application of science principles in processes, tools, and devices.

tectonics: The change in a surface of a planet due to internal forces.

terrestrial: Of or having to do with solid rock; name given to the four inner planets (Mercury, Venus, Earth, and Mars). See also *planet.*

universe: The entirety of everything that is known to exist in space.

velocity: Speed and direction that an object travels over a specified distance during a measured amount of time; rate of motion.

volcano: A landform, usually cone shaped, produced by a collection of erupted material around a vent, or opening, in the surface of a planet or moon and through which gas and erupted material pass.

weight: A measure of the force of gravity on an object.

year: The time it takes a planet to complete one revolution around the Sun.

Index